D0919348

WITHDRAWN
No longer the property of the
Boston Public Library.
Sale of this material benefits the Library

# The Chernobyl Nuclear Disaster

GREAT DISASTERS
REFORMS and RAMIFICATIONS

# The Chernobyl Nuclear Disaster

Kristine Brennan

CHELSEA HOUSE PUBLISHERS
Philadelphia

*Frontispiece:* This exhibit shows some of the protective masks and other gear workers used to protect themselves from radiation during the cleanup of the area around the Chernobyl plant. As the worst peacetime nuclear disaster in history, the Chernobyl disaster stands as a costly reminder of the dangers of nuclear power.

**CHELSEA HOUSE PUBLISHERS**

*Editor in Chief* Sally Cheney
*Director of Production* Kim Shinners
*Production Manager* Pamela Loos
*Art Director* Sara Davis
*Production Editor* Diann Grasse

**Staff for THE CHERNOBYL NUCLEAR DISASTER**

*Editor* Bill Conn
*Associate Art Director* Takeshi Takahashi
*Layout* 21st Century Publishing and Communications, Inc.

© 2002 by Chelsea House Publishers, a subsidiary of Haights Cross Communications. All rights reserved.
Printed and bound in the United States of America.

First Printing

1 3 5 7 9 8 6 4 2

The Chelsea House World Wide Web address is
**http://www.chelseahouse.com**

Library of Congress Cataloging-in-Publication Data

Brennan, Kristine, 1969–
The Chernobyl nuclear disaster / by Kristine Brennan.
p. cm — (Great disasters, reforms and ramifications)
ISBN 0-7910-6322-4 (alk. paper)
1. Chernobyl Nuclear Accident, Chernobyl§' Ukraine, 1986. I. Title. II. Series.

TK1362.U38 B74 2001
363.17'99'0947714—dc21

2001047324

YA
TK1362
.V38B74
2002

# Contents

**GREAT DISASTERS**
**REFORMS and RAMIFICATIONS**

THE *APOLLO I* AND *CHALLENGER* DISASTERS

BHOPAL

THE BLACK DEATH

THE BLIZZARD OF 1888

THE BOMBING OF HIROSHIMA

THE CHERNOBYL NUCLEAR DISASTER

THE DUST BOWL

THE EXPLOSION OF TWA FLIGHT 800

THE *EXXON VALDEZ*

THE GALVESTON HURRICANE

THE GREAT CHICAGO FIRE

THE GREAT PLAGUE AND FIRE OF LONDON

THE *HINDENBURG*

THE HOLOCAUST

THE INFLUENZA PANDEMIC OF 1918

THE IRISH POTATO FAMINE

THE JOHNSTOWN FLOOD

LOVE CANAL

THE MUNICH OLYMPICS

NUCLEAR SUBMARINE DISASTERS

THE OKLAHOMA CITY BOMBING

PEARL HARBOR

THE SALEM WITCH TRIALS

THE SAN FRANCISCO EARTHQUAKE OF 1906

THE SPANISH INQUISITION

THE STOCK MARKET CRASH OF 1929

TERRORISM

THREE MILE ISLAND

THE *TITANIC*

THE TRIANGLE SHIRTWAIST COMPANY FIRE OF 1911

THE WACO SIEGE

THE WORLD TRADE CENTER BOMBING

**Jill McCaffrey**
National Chairman
Armed Forces Emergency Services
American Red Cross

# Introduction

Disasters have always been a source of fascination and awe. Tales of a great flood that nearly wipes out all life are among humanity's oldest recorded stories, dating at least from the second millennium B.C., and they appear in cultures from the Middle East to the Arctic Circle to the southernmost tip of South America and the islands of Polynesia. Typically gods are at the center of these ancient disaster tales—which is perhaps not too surprising, given the fact that the tales originated during a time when human beings were at the mercy of natural forces they did not understand.

To a great extent, we still are at the mercy of nature, as anyone who reads the newspapers or watches nightly news broadcasts can attest.

Hurricanes, earthquakes, tornados, wildfires, and floods continue to exact a heavy toll in suffering and death, despite our considerable knowledge of the workings of the physical world. If science has offered only limited protection from the consequences of natural disasters, it has in no way diminished our fascination with them. Perhaps that's because the scale and power of natural disasters force us as individuals to confront our relatively insignificant place in the physical world and remind us of the fragility and transience of our lives. Perhaps it's because we can imagine ourselves in the midst of dire circumstances and wonder how we would respond. Perhaps it's because disasters seem to bring out the best and worst instincts of humanity: altruism and selfishness, courage and cowardice, generosity and greed.

As one of the national chairmen of the American Red Cross, a humanitarian organization that provides relief for victims of disasters, I have had the privilege of seeing some of humanity's best instincts. I have witnessed communities pulling together in the face of trauma; I have seen thousands of people answer the call to help total strangers in their time of need.

Of course, helping victims after a tragedy is not the only way, or even the best way, to deal with disaster. In many cases planning and preparation can minimize damage and loss of life—or even avoid a disaster entirely. For, as history repeatedly shows, many disasters are caused not by nature but by human folly, shortsightedness, and unethical conduct. For example, when a land developer wanted to create a lake for his exclusive resort club in Pennsylvania's Allegheny Mountains in 1880, he ignored expert warnings and cut corners in reconstructing an earthen dam. On May 31, 1889, the dam gave way, unleashing 20 million tons of water on the towns below. The Johnstown Flood, the deadliest in American history, claimed more than 2,200 lives. Greed and negligence would figure prominently in the Triangle Shirtwaist Company fire in 1911. Deplorable conditions in the garment sweatshop, along with a failure to give any thought to the safety of workers, led to the tragic deaths of 146 persons. Technology outstripped wisdom only a year later, when the designers of the luxury liner *Titanic* smugly declared their state-of-the-art ship "unsinkable," seeing no need to provide lifeboat capacity for everyone onboard. On the night of April 14, 1912, more than 1,500 passengers and crew paid for this hubris with their lives after the ship collided with an iceberg and sank. But human catastrophes aren't always the unforeseen consequences of carelessness or folly. In the 1940s the leaders of Nazi Germany purposefully and systematically set out to exterminate all Jews, along with Gypsies, homosexuals, the mentally ill, and other so-called undesirables. More recently terrorists have targeted random members of society, blowing up airplanes and buildings in an effort to advance their political agendas.

The books in the GREAT DISASTERS: REFORMS AND RAMIFICATIONS series examine these and other famous disasters, natural and human made. They explain the causes of the disasters, describe in detail how events unfolded, and paint vivid portraits of the people caught up in dangerous circumstances. But these books are more than just accounts of what happened to whom and why. For they place the disasters in historical perspective, showing how people's attitudes and actions changed and detailing the steps society took in the wake of each calamity. And in the end, the most important lesson we can learn from any disaster—as well as the most fitting tribute to those who suffered and died—is how to avoid a repeat in the future.

# Fire in the Sky

An explosion and fire in the Chernobyl-4 reactor initiated a series of catastrophes in 1986. Chernobyl plant supervisors, unaware of the severity of the situation, failed to notify the community of the radioactive material emanating from the plant until it was too late.

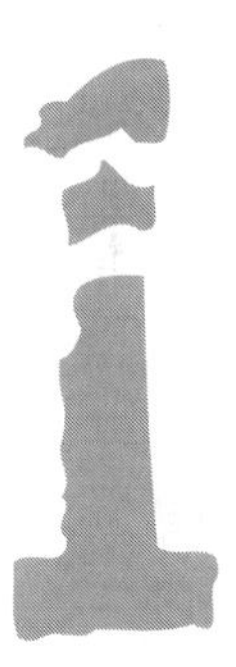

On the morning of April 26, 1986, many residents of Pripyat, a town in the Ukrainian republic of the Soviet Union with a population of 49,360, stepped onto their balconies to greet the warm spring sun. Little did they know that the fresh air was highly *radioactive*—containing invisible, dangerous particles capable of seriously sickening or killing animals and people. Although smoke and fire were clearly visible from the V. I. Lenin NPP (Nuclear Power Plant), just three kilometers (km) outside of Pripyat, nobody was especially alarmed. After all, the power plant—commonly called Chernobyl after a nearby industrial city—was known to be extremely safe. Many of Pripyat's residents worked there. They were certain firefighters would quickly contain this minor blaze. Moreover, nobody had been notified of any potential danger. School was open, and the day promised to be a nice one for running errands and preparing for May Day, an upcoming labor holiday.

Pripyat's residents couldn't have been more wrong. At 1:23:58 A.M., at least two powerful explosions had ripped open the fourth reactor unit at the plant, called Chernobyl-4. The distant blasts awakened a few people, but they soon went back to sleep. But by the time early risers in Pripyat were rolling out of bed, the accident had already snatched two lives. Valery Khodemchuk, a plant worker who had been monitoring the pumps that moved water in and out of the reactor core, died first when the explosions hurled Chernobyl-4's 1000-ton concrete lid into the air. The enormous shield, meant to keep radioactive material inside the reactor, flipped over once and landed at an angle, partially covering the reactor. Burning radioactive material shot out of the uncorked reactor. Hot radioactive gases and particles formed a deadly cloud that would cut a terrifying path over the USSR and Europe for days. Much of the hot debris fell back to earth, touching off numerous fires in and around the building that housed Chernobyl-4. Valery Khodemchuk was probably boiled alive by steam in the first blast. A pile of highly irradiated rubble then rained down on him, permanently entombing his body.

Khodemchuk's supervisor, Valery Perevozchenko, made a valiant effort to find him. Perevozchenko called for Khodemchuk as loudly as his burning lungs and throat permitted, rummaging desperately through the contaminated wreckage. The supervisor searched in vain for Khodemchuk until his already lethal dose of *radiation* left him too ill to continue. Perevozchenko would join Khodemchuk in death before long.

The lights in Chernobyl-4 went out, forcing workers with only their pocket flashlights to navigate a nightmare world of fire, twisted metal, ruptured pumps, and gushing water. Two coworkers pulled an unconscious Vladimir Shashenock from the wreckage and carried him to safety. One of the rescuers later got radiation burns on his back where Shashenock's limp hand had rested. Shashenock's body was literally emitting radiation, although his internal injuries and thermal burns killed him before the radiation could. He would remain unconscious until his death at around 6:00 A.M. in the Pripyat medical center.

The heroism of workers in Unit Four prevented an already unimaginable catastrophe from worsening. Machinists helped fight fires in the turbine hall. Razim Davletbayev, deputy chief of Chernobyl-4's turbines, coordinated the electricians' efforts to remove volatile hydrogen from the reactor's generators and replace it with more stable liquid nitrogen. He also directed them to get the emergency oil reserves for the backup diesel generators away from the fires. His quick thinking—and the courageous actions of the electricians—spared the plant's three remaining reactor units by limiting the spread of the fires.

The selflessness of Aleksandr Lelechenko, deputy chief of the electricians, cost him his life. He repeatedly entered a highly contaminated room in Chernobyl-4 to cut off the hydrogen supply to the backup generators. Lelechenko assumed the lion's share of the risk to protect his subordinates from exposure to a pile of radioactive debris from the explosions. Before he died, doctors estimated that Lelechenko had absorbed enough radiation to kill several men.

In the chaotic first hours after the explosions, plant workers had no idea how much radiation they were absorbing. Although they were equipped with *dosimeters,* devices that measured personal radiation levels, the

This Russian-language map shows the region of the Ukraine and Belarus that in 1986 suffered the most immediate damage from nuclear fallout. The Chernobyl plant, named after a nearby city, is actually located in the city of Pripyat.

instruments were designed only to monitor lower doses of radiation. They only went up to about four *roentgens* per hour. Later, Grigory Medvedev, author of the book *The Truth About Chernobyl*, estimated that throughout the reactor building, radiation emissions ranged from roughly 1,000 to 20,000 roentgens per hour! Radiation doses absorbed by living tissue are expressed in rads (one rad equals roughly one roentgen), which are in turn converted to equivalent rems (*r*oentgen *e*quivalent *m*an) to measure the effects of a given dose on the human body. To put the Chernobyl disaster into perspective, incidental radiation exposure from solar and cosmic rays, rocks and soil, and X-ray treatments totals less than 400 millirems *per year* for most Americans (.4 rems). The radiation dose *per hour* at Chernobyl-4 was at least 2,500 times that amount.

Dosimeters able to register higher radiation levels would arguably have saved lives and helped preserve the future health of many people by warning them to leave the vicinity of the plant immediately. Better dosimetric equipment was on the premises, but it was locked inside a safe—and nobody present had a key! But such unpreparedness was consistent with the Soviet nuclear power program's habitual denial that emergencies were possible. This pervasive denial led workers to underestimate the hazards they faced and to overestimate the safety of the nuclear power plants in which they worked. Like the proverbial tree falling in a forest, an accident in a Soviet NPP did not exist unless plant directors and operators saw it for themselves. Even then, the Communist Party forced supervisors to suppress the information under threat of punishment. Though technically just a political party, the Communist Party had complete control of the government of the Soviet Union. This one-party system produced a highly bureaucratic dictatorship. Until April 26, 1986, the Chernobyl plant had been one of the crown jewels of the Soviet Union's nuclear power industry. Its employees steadfastly believed the Communist Party's affirmations that the facility was virtually accident-proof.

The inability to measure true levels of radioactivity at the time of the accident probably helped solidify an irrational belief held by the plant's top officers: they insisted that the destroyed reactor was still whole. Graphite, normally present only inside the reactor core, lay in pieces both inside and outside of the building. People were rapidly becoming nauseous, faint, and strangely excited. Anyone at all familiar with the plant and with the symptoms of radiation poisoning should have immediately recognized that the reactor was damaged and that ambient radiation levels were much higher than normal. But on this fateful morning, rational thought yielded to panic, disbelief, and the fear of being held personally responsible for the accident.

Anatoly Stepanovich Dyatlov, deputy chief operational engineer for Units Three and Four, and shift foreman Aleksandr Akimov repeatedly told anyone who would listen that Reactor Four was intact. Clinging to this belief allowed them to hope that they could contain the accident. Dyatlov and Akimov couldn't see the destroyed reactor from their posts in the control room. Akimov sent two hapless young engineer trainees to where the reactor should have been. He told them to manually lower the *control rods* into the core to shut it down. Normally, control rods were used to keep the nuclear reaction inside the core in check. The more control rods inserted into the core, the slower the nuclear reaction. When the trainees returned to tell Akimov and Dyatlov that there was no reactor left to shut down, their report fell on deaf ears. Death from radiation poisoning would be the young men's reward for attempting to carry out Akimov's orders.

Valery Perevozchenko, reeling from his brave but fruitless search for the buried Valery Khodemchuk, also told Akimov that the reactor had blown up. Akimov insisted that the reactor was still intact and that it needed water.

Why did Aleksandr Akimov systematically ignore anyone who told him the truth about Reactor Four? At this stage of the accident, it was still possible to deny the full horror of what had happened. Only two people had died so far. Nobody had gotten accurate readings of the radioactivity levels in the air—although the top men at

After the explosion, the area near Reactor No. 4 at the Chernobyl plant was strewn with the rubble of twisted girders and broken concrete beams. The real danger, however, was not the initial explosion but the radiation that was released from the reactor into the environment.

the plant became defensive and angry whenever skilled dosimetrists warned them that the radiation fields were anything but normal. Akimov was certain he had done everything right up to the moment of the accident. If Party officials in Moscow were not aware of a problem at the plant, then no problem existed. As Akimov's denial continued to snowball, several men walked to their deaths, checking the status of a reactor that was obviously destroyed. This stubborn denial of danger soon spread to the very top level of the plant's management. When plant director Viktor Petrovich Bryukhanov arrived at 2:30 A.M., Akimov told him that Reactor Four was intact. In an administrative building not far from Chernobyl-4, Bryukhanov was on the phone to officials in Moscow by 3:00 A.M., explaining that the reactor was still in one piece and that radiation levels were within normal limits.

Chief engineer Nickolai Masimovich Fomin arrived at 4:30 A.M. When Akimov told him that the reactor was intact, Fomin ordered him to feed it water. To implement this command, Fomin dispatched Akimov, senior reactor control engineer Leonid Toptunov, and Anatoly Dyatlov to go around Unit Four and manually open valves that would send water into the reactor core. Akimov and Toptunov labored mightily, all the while sealing their doom from radiation poisoning. They gave their lives acting on Fomin's insane hope of saving a reactor that was already a fiery crater—an insane hope that they still shared.

The irrational notion that the reactor was salvageable persisted into late morning. Fomin asked physicist Anatoly Sitnikov to go look at Reactor Four. At around 10:00 A.M. Sitnikov returned with the same news as those preceding him: the reactor was destroyed. Fomin's inability to face the truth would cost Anatoly Sitnikov his life in the coming weeks.

The grim truth about Chernobyl-4—a truth that several doomed individuals had already tried to tell Bryukhanov, Fomin, Akimov, and Dyatlov—was even more painfully obvious from overhead. Boris Prushinsky, chief engineer of the Soviet department of nuclear energy, surveyed the blazing reactor from a helicopter. After Prushinsky saw Chernobyl-4, the charade was over. Bryukhanov and Fomin could no longer believe the myth that everything was under control. Bryukhanov would spend the next several days "hanging around, like a piece of [expletive]," as he put it, while a panel of experts assembled in Pripyat to search for ways to manage the disaster he had attempted to cover up.

As dozens of fires burned in and around the plant, fire and medical personnel mobilized from Pripyat. Six firefighters were on duty at the plant when the fires started, but additional patrols hurried to the scene. Major Leonid P. Telyatnikov, commander of the fire brigade serving the nuclear power plant, hurried in to coordinate the firefighting effort. Vladimir Pravik, leader of the Number Two Patrol, directed all fire engines in Ukraine's Kiev region to the stricken power plant. Specialized Military Fire Brigade Number Six was the first unit to climb to the roof of the turbine hall, where pieces of highly radioactive graphite had landed after the blasts.

It was not long before the men of Brigade Number Six were too sick to continue. The firefighters became faint, nauseous, and weak. The next wave of firefighters helped their shaky colleagues leave the roof, then continued to battle the fire on top of the turbine hall until it was finally out at around 5:00 A.M. Firefighters would account for many of the 29 Chernobyl-related deaths at Moscow Hospital No. Six in the months following the disaster. The surviving firefighters returned to work after months of rehabilitation—and then only to desk jobs. The lung damage they suffered at Chernobyl left them unfit for active duty. By 1989, all of the firefighters who had been on the roof of the turbine hall were dead except for Major Telyatnikov.

The combined fire-fighting patrols extinguished some 37 fires around Reactor Four. But the burning reactor itself eluded their efforts. Fruitless attempts to douse the inferno with water continued for hours. The firefighters didn't know that water is useless against a nuclear fire.

As the firefighters battled flames on top of the turbine hall, the first ripples of the accident were making themselves felt just outside of the devastated nuclear power

Family members of a firefighter killed in action grieve near his tomb, 15 years after the 1986 explosion. A number of firefighters along with Chernobyl plant workers lost their lives in the disaster.

plant. When a 28-year-old doctor named Valentin Belokon went on duty at the Pripyat medical center on the evening of April 25, he could not have imagined that he would be a hero by morning. His assistant, Aleksandr Skachok, was already out at the plant with an ambulance when he called for Dr. Belokon at around 1:40 A.M.

Skachok returned to Pripyat in his ambulance with the dying Vladimir Shashenock.

As the first physician to respond to the accident, Belokon found himself inundated with people in need of treatment. His patients' complaints included severe nausea and faintness and painfully burning eyes and throats. Many of the worst victims had "nuclear tans" that had toasted their faces and limbs to a dark, swollen brown. It was all Belokon could do to sort the sickest from the sick and send the most urgent cases to the hospital first. The firefighters, although very weak, were utterly unable to calm down after their ordeal. Dr. Belokon could do little more than administer tranquilizers to them. Author Grigory Medvedev later described their marked agitation as "a nuclear frenzy of the nervous system."

At around 6:00 A.M., Valentin Belokon finally fell to radiation sickness that ultimately landed him in Moscow's Hospital No. Six. Although Dr. Belokon recovered and went on to specialize in pediatric medicine, his swift response to the accident at Chernobyl left him with respiratory problems.

The immediate medical problems that Dr. Belokon observed were just the beginning. The accident would affect more land, water, and air than anyone could initially imagine. Yet the Soviet Union behaved as if the radioactive cloud rising from Chernobyl could be tidily contained within its own borders. The story of Chernobyl is a story of overconfidence, unpreparedness, and disregard by the highest levels of the Communist Party for the safety of both the Soviet people and those of neighboring nations. It is also the story of an inherently dangerous type of nuclear reactor and a blundered experiment that led to a global disaster.

ACCURACY
ENHANCING
MODE
POCKET SURVEY METER
19.99
µSv/h
POWER
OFF ON
1850-338
MYRATE
PDR-101

A researcher at the Chernobyl nuclear plant holds up a Geiger counter used to measure radiation levels in the air. Many Chernobyl plant operators had similar equipment on the day of the accident, though their devices failed to record the high levels of radiation emanating from the reactor core.

# Dangerous Design, Deadly Experiment

2

The warm, beautiful Saturday morning of April 26, 1986, had turned into a nightmare of unleashed atoms at Chernobyl. To understand how and why things went so terribly wrong, one must first understand a little bit about atomic energy and about the nuclear power reactors at Chernobyl.

All active commercial nuclear power reactors produce electricity using heat from the *fission*, or splitting, of the *nuclei* of atoms. Certain types of elements can be split in this way when they are bombarded with *neutrons*, a type of particle found in the nucleus of nearly all atoms. Uranium and plutonium are two elements that are commonly used in nuclear reactions because a certain type, or *isotope*, of these elements is easily susceptible to fission. The reactors at Chernobyl were fueled by uranium, less than two percent of which was the fissionable isotope uranium 235. In other words, the uranium in the plant was enriched to less than two percent uranium 235; the rest was the more common and non-fissile uranium 238. Because it is *unstable*, an atom of uranium 235 ejects a neutron from its nucleus when it is hit by another neutron. This occurs spontaneously, and a huge amount of energy is released when it happens.

To keep fission going in a *chain reaction,* a sufficient number, or critical mass, of uranium 235 atoms must be close enough together. Each liberated neutron will collide with another atom of uranium 235, splitting its nucleus, freeing another neutron, and releasing more energy. The newly released neutrons will continue the fission reaction by splitting more uranium 235 nuclei.

To slow the neutrons to the right speed for a sustained, controlled chain reaction that would generate only heat energy, the reactors at Chernobyl had *moderator* blocks made from graphite, a form of carbon. The heat from the nuclear reaction boiled water in the reactor core into steam. The steam pressure drove turbines located outside the core, generating electricity. The four reactors at Chernobyl (Units Five and Six were under construction at the time of the accident) each drove two turbogenerators. Reactor Four drove Turbogenerators Seven and Eight.

The nuclear fuel inside the Chernobyl reactors was in the form of uranium oxide pellets stacked into tubes a little more than 3.5 meters long. Thirty-six of these tubes, called fuel rods, were bundled together to form fuel assemblies about 10 meters long. The graphite moderator blocks were stacked between these fuel assemblies to regulate the speed of the neutrons. Chernobyl-4 had about 2,500 graphite columns housed between some 1,700 fuel assemblies.

When a nuclear chain reaction is unchecked, it can lead to an explosion. The atomic bomb at Hiroshima owed its destructive power to the runaway fission of uranium 235. But Chernobyl's reactors also had 211 control rods made of boron carbide (a neutron-absorbing material) to slow or even stop fission, depending upon the number of rods lowered into the reactor core. The instruction manual for operators at the plant specified that a minimum of 30 control rods should be fully lowered into the core at all times for maximum safety.

The type of reactor used at Chernobyl is called an RBMK reactor. The Soviet Union originally designed the RBMK reactors to be breeder reactors, which actually produce more fissile material than they need to sustain their chain reactions. In breeder reactors, some of the non-fissile uranium 238 present in the fuel rods is transformed into plutonium 239, a fissile isotope that can be used to make nuclear weapons. Although Chernobyl-4 and the Soviet Union's other RBMKs built to generate electricity were ultimately not used as breeder reactors, they still produced some plutonium 239 during the fission of the uranium fuel. Over time, the sustained fission reaction produced many other *radionuclides*, or radioactive atoms, inside the reactor core as well. The metal coating around the fuel rods absorbed these fission products as they accumulated.

The RBMK reactors had a number of safety issues. Unlike most other reactors, which rely on water to both cool the core and to moderate fission, RBMKs are graphite-moderated. But graphite is extremely flammable, making the RBMK more dangerous than water-moderated reactors in the event of accidents. Another dubious feature of the RBMK is a relatively weak containment vessel surrounding the core. A thick concrete containment building helps prevent the escape

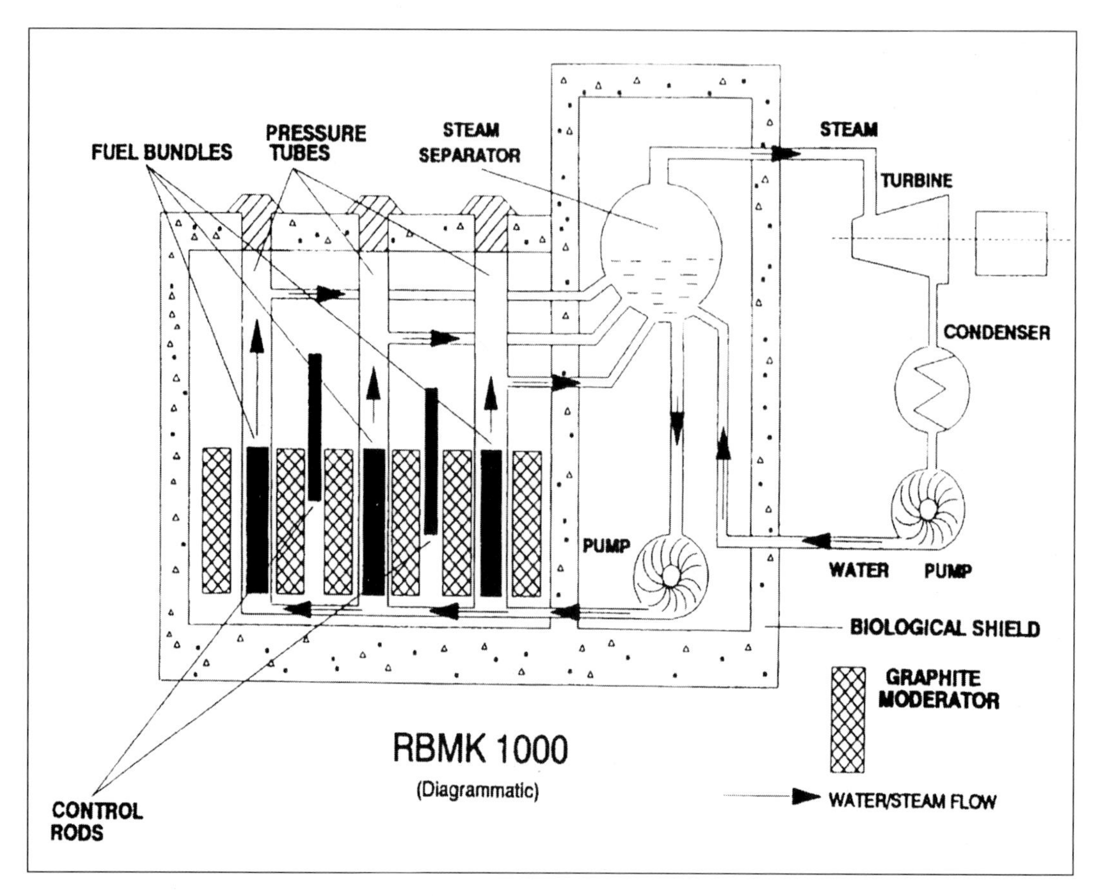


This diagram of the RBMK reactor shows how its nuclear reactions produce energy in the form of steam. Unlike most reactors, which shut down after a buildup of excess steam pressure, the RBMK reactor allows reactivity to increase to dangerous levels with the rise of steam pressure, as occurred in the Chernobyl-4 explosion.

of dangerous gamma radiation into the environment (alpha and beta radiation are not as penetrating). The RBMK lacks such a thick-walled containment building. This deficiency also leaves RBMK-type reactors extremely vulnerable to blowing apart when excess steam pressure builds up inside them (as happened at Chernobyl).

The RBMK reactor's most dangerous characteristic, however, is its positive void coefficient. The void coefficient indicates how much steam the nuclear core is producing. Reactors that use water to both cool and moderate the reaction have a negative void coefficient. This means that excess steam production in the core will eventually shut the reactor down instead of boosting its power.

The RBMK is the only type of commercial reactor with a positive void coefficient. In this design, if too much water boils off into steam, or if coolant water gets displaced from its channels inside the reactor, the chain reaction increases, which leads to increased power output and still more steam production. The additional steam increases reactivity even more. The RBMK's positive void coefficient carries with it a risk of uncontrolled fission and *meltdown* inside the core. What the RBMK's designers counted on, however, was that once nuclear fuel would heat up to a certain temperature, its reactivity would start to decrease. This decrease in reactivity would slow steam production, increasing the amount of coolant water available to the reactor. As long as an RBMK reactor operates at an output level of 700 thermal megawatts or higher, the tendency of nuclear fuel to become less reactive at very high temperatures helps counteract the hazards of a positive void coefficient. Unfortunately, at lower power output levels the RBMK reactor is very unstable and prone to sudden power surges.

At the time of the accident, the RBMK reactor was an important facet of the Soviet Union's nuclear energy program. Reactor Four at Chernobyl was due for a tune-up on April 25, 1986. Online since its December 1983 start-up, it was about to be shut down for scheduled inspections and maintenance. Technicians would replace numerous worn-out fuel assemblies during this maintenance period. The fission products that build up inside the fuel-rod coating poison the reactor, dampening its reactivity and making power distribution uneven throughout the core. By the time of the accident, numerous radionuclides had accumulated inside Chernobyl-4. The disaster helped to release these particles into the atmosphere along with hot nuclear fuel.

Plant director Bryukhanov and chief engineer Fomin planned to run a test in the reactor unit on April 25 before the shutdown, despite the buildup of radionuclides in the core. They wanted to test the performance of a magnetic device installed inside Turbogenerator Eight. Nuclear power plants have backup generators to provide auxiliary electricity for vital protection systems such as control rods and cooling water pumps during blackouts. But the diesel backup generators at Chernobyl took one minute or more to power up—long enough for core melting to start without available coolant or operational control rods. The magnetic device in Turbogenerator Eight was supposed to ensure that if the steam pressure from the reactor was cut off, the turbines, spinning on inertia alone, could generate enough electricity to support the reactor until the backup generators were ready to take over.

The test itself was not new. It had been conducted previously on Chernobyl's other reactors and at other NPPs. But in the case of Chernobyl-4, the tests should have already been completed. Reactor Four had gone on-line on December 20, 1983, in time to coincide with a Soviet labor holiday honoring the energy industry on December 22. Appearance was everything in the Soviet Union: the façade of accident-proof workplaces and of workers who typically beat production deadlines was maintained at all costs. This meant that Reactor Four's turbogenerator tests were left incomplete to rush its start-up date. Before a reactor went on-line, it was supposed to undergo a whole battery of tests through a process called verification. But parts of Chernobyl-4's verification were skipped altogether to meet a deadline imposed to make the energy industry appear efficient.

Bryukhanov and Fomin's test plan was equally slipshod. It included instructions to shut down the emergency core cooling system (ECCS) during the test. This system sent coolant water to the reactor whenever it sensed that too much was boiling off into steam. Why would Bryukhanov and Fomin elect to disable such an important safety feature? The plant director and the chief engineer wanted to avoid flooding the reactor with water so that they could repeat the test immediately if necessary. Bryukhanov and Fomin had submitted a written test plan to the Ukraine bureau of the Ministry of Power and Electrification (Gidroproyekt) back in January of that year. Although they never got a reply from the bureau, they decided to proceed anyway.

Gidroproyekt should have immediately denied permission for Bryukhanov and Fomin's proposed test. Their plan clearly stated their intention to disable the ECCS—a practice unheard of, even during tests. Other parts of the test plan were also poorly constructed, as leading Soviet scientist Valery Legasov would later reveal in writings published after the accident.

Bryukhanov and Fomin's dangerous and convoluted test plan showed that both men had a surprisingly limited understanding of the intricacies of nuclear reactors. Viktor Petrovich Bryukhanov, director of the Chernobyl NPP, was a turbine specialist with no formal education in atomic energy. Nuclear expert and author Grigory Medvedev, who worked with Bryukhanov in the early 1970s, credited him with intelligence and a good work ethic. Even so, Medvedev also noted, Bryukhanov's strong points hardly qualified him for the top job at a nuclear power plant. Nuclear power plant management was a prestigious career in the Soviet energy industry. Therefore, administrative posts at nuclear plants were sometimes handed out to reward loyal Party members who had risen through the ranks at traditional coal-burning power plants. As Bryukhanov advanced in the nuclear power industry, he often nominated people with weak technical backgrounds in atomic energy for key plant management positions.

In 1972, Bryukhanov nominated Nickolai Masimovich Fomin to be chief of Chernobyl's electrical unit. Fomin, an electrical engineer, was promoted to chief engineer of the entire plant despite the Ministry of Energy's recommendation of another candidate with a strong background in atomic energy. But Bryukhanov pushed for Fomin. Eventually, local Communist Party pressure got Fomin installed as chief engineer and deputy director of the plant.

Less than a year before the accident at Chernobyl, Fomin had suffered severe injuries in a car accident. His injuries had been so debilitating that he had been back to work for just one month when the Chernobyl accident occurred. Grigory Medvedev was inspecting construction on the unfinished Chernobyl-5 at the time of Fomin's return. He described the tense and edgy chief engineer as "a shadow of his former self," saying, "His whole being had clearly slowed down and bore the imprint of the pain he had endured." If Medvedev's assessment is accurate, then Fomin may not have been in any shape to make sound judgments when disaster struck on the morning of April 26.

Confident despite their limited expertise, Bryukhanov and Fomin went ahead with their unauthorized experiment. It was scheduled to begin on the day shift at 1:00 P.M. on April 25. At 1:05, Turbogenerator Seven was shut down to facilitate testing of Turbogenerator Eight. At 2:00 P.M., the ECCS was disabled. Shortly thereafter, a Ukrainian electrical official instructed plant personnel to delay the experiment until 11:00 P.M., when consumer demand for electricity would be

lower. This was an important consideration since Chernobyl supplied Ukraine with roughly 45% of its nuclear-generated electricity.

Chernobyl-4 ran without its ECCS throughout the 10-hour delay. The delay of the test meant that the night shift—which began at midnight—would be running the experiment. The day shift operators were not only more experienced than the night operators, but they were also the only ones who had received any briefing on the test program, flawed as it was. The night shift included shift supervisor

Plant director Bryukhanov and chief engineer Fomin made many fatal mistakes, including disabling the emergency core cooling system, during an unauthorized experiment on April 25 and 26. Here, two workers check themselves for radiation after leaving the sarcophagus that entombs reactor No. 4.

Anatoly Dyatlov and the young senior reactor control engineer Leonid Toptunov. These two individuals were destined to become key players in the events leading up to the accident.

Disaster was fast approaching at 12:28 A.M. on April 26. The operators had begun lowering the power level of Chernobyl-4 to 700 thermal megawatts. The power level required for the test to yield meaningful results was 700–1,000 thermal megawatts. Unfortunately, the power level continued to drop. Expert Richard Mould, author of *Chernobyl Record,* says that reactor power was only about 500 megawatts at this time. An RBMK reactor becomes difficult to control at levels lower than 700 megawatts. At some point around this time, one of the operators mistakenly disabled the local automatic control rods, a protection system that regulated reactor power at higher levels. This error made it impossible for the operators to get the reactor back to the baseline level of 700 megawatts.

The test was already doomed to failure. The power level in Chernobyl-4 soon collapsed to 30 megawatts. When power drops this low, the risk of reactor poisoning increases. Anatoly Dyatlov, the deputy chief engineer, should have ordered Akimov and Toptunov to halt the test at this point. Instead, Dyatlov cursed loudly and shouted at his subordinates to raise the reactor's power.

The only way to increase power to the reactor was to remove control rods from its core. Standard operation of the reactor called for a minimum of 30 rods to be fully inserted at all times; there were only 28 at this stage of the test. Taking out more was a gamble that Leonid Toptunov hesitated to take. He initially defied Dyatlov's orders to do so. But the 26-year-old senior reactor control engineer soon caved in to his supervisor's continued yelling and removed more control rods. Grigory Medvedev wrote:

> Toptunov began to increase power, thereby signing a death sentence for himself and for many other comrades. That symbolic sentence also carried the clearly visible signatures of Dyatlov and Fomin. Other discernible signatures included those of Bryukhanov and many other more highly placed comrades.

Reactor power began to increase, but the operators' ability to control the reactor began to diminish. Attempts to reconstruct the conditions at the time of the accident have shown that about eight control rods were in place at the time of the explosion. With so few rods inserted, the reactor's emergency power reduction system was no longer able to halt fission inside the reactor if necessary.

At around 1:00 A.M., Toptunov and Akimov had raised reactor power to 200 megawatts—still too low for the experiment to be useful—but they were under orders to proceed. For a brief time, the reactor was stable. But at such a low power level, Chernobyl-4 was steadily accumulating poisons that threatened to destabilize it. Toptunov again attempted to increase the power level. He switched on more coolant pumps so that the reactor could produce more steam, hoping that its intrinsic positive void coefficient would quickly drive up the power. Initially, the flood of additional water actually decreased reactivity. This prompted Toptunov to pull out more control rods. To avoid disrupting the experiment, the operators had disabled the last safety feature that could have saved Chernobyl-4: a system that would shut down the reactor upon sensing that two turbines had shut down (Turbogenerator Seven had been turned off since the afternoon of April 25.).

Steam pressure was cut off to Turbogenerator Eight at 1:23 A.M. Less than one minute remained until the worst peacetime nuclear disaster in history. The turbine continued spinning on inertia, in accordance with the test plan.

Soon, the reactor was losing coolant water as it boiled away into steam. Because of the RBMK's positive void coefficient, the power level began to increase. Fearing a power surge, Akimov (or Toptunov, according to some accounts) allegedly pressed an emergency button to lower all 211 control rods into the core at 1:23:40 A.M. But the control rods needed 18 seconds to fully insert into the core. Although the main body of each control rod was made of boron to absorb neutrons and stop fission, the rods had graphite tips and hollow sections one meter long immediately behind the tips. The tips and the hollow parts of the rods displaced water when first inserted, and they momentarily *increased* reactivity when they entered the core. In normal operating situations, this increase was brief and inconsequential. This time, however, in a matter of seconds the lowering of the rods skyrocketed the unstable reactor's power to many times the normal level.

Operators all over the reactor building could feel the floor shaking. A terrified Akimov tried to cut off the motor that lowered the control rods, hoping that gravity would drop them the rest of the way into the core. It was too late, however.

The power surge led to drastically increased steam pressure inside Chernobyl-4. Superheated steam exploded, followed by what was probably a hydrogen explosion. Coolant channels were twisted and mangled. The inadequate containment vessel around the reactor blew open. The 1000–metric ton shield on top of the building shot into the air, landing wedged at a nearly vertical angle over the burning crater that had been a reactor just a moment earlier. The graphite left inside Chernobyl-4 was in flames, having reacted with oxygen after the first explosion.

Broken fuel assemblies and blazing graphite moderator blocks spewed forth, touching off dozens of fires in and around the building. As the catastrophe began to unfold further, the USSR embarked on a futile effort to hide the truth.

Civilians in the area surrounding Chernobyl received short notice of their evacuation, as this picture of a neighborhood in Solnechny, Belarus, shows. Many were under the misguided impression that they would return home in a few days and so left their essential belongings behind.

# 3 Time Is of the Essence

It wasn't until about 9:00 P.M. on April 26 that a government commission convened at Party headquarters in Pripyat to cope with the effects of the accident. It was a Saturday and a holiday weekend in honor of May Day. Some officials named to the commission had been vacationing when the accident happened. Valery Legasov was present; he was the deputy director of the USSR's most prestigious nuclear research entity, Moscow's Kurchatov Institute of Atomic Energy. Legasov was standing in for the official director, Anatoly Alexandrov, considered the father of the RBMK reactor. Boris Yevdokimovich Shcherbina, deputy chair of the Soviet Union's Council of Ministers, chaired this special meeting in Pripyat. Also in attendance was the Minister of Energy and Electrification, Anatoly Ivanovich Mayorets, an expert on electrical transformers.

Chernobyl plant director Viktor Petrovich Bryukhanov and chief engineer Nickolai Masimovich Fomin were there, too. Both men would see the wreckage of Chernobyl-4 firsthand from a helicopter window. Bryukhanov would finally admit that Chernobyl-4 was indeed destroyed; Fomin would soon suffer a nervous breakdown.

As the chairman of the panel, only Boris Shcherbina was empowered to order the evacuation of Pripyat. He made it clear, however, that he was unwilling to do so. He dreaded causing a panic among the people far more than he feared widespread radiation. The commission's radiation experts urged him to change his mind. Every second that he and Minister Mayorets continued to ignore the scientists' pleas meant more radiation exposure for the people of Pripyat. Author Grigory Medvedev wrote:

> He [Shcherbina] still did not realize that the air around him—in the street and inside the room—was saturated with radioactivity, and emitting gamma and beta rays which penetrated whomever happened to be in their way—ordinary mortals, Shcherbina, or the devil himself. As for ordinary mortals, there were about 48,000 of them in the town that night, including senior citizens, women, and children.

In addition to trying to convince Shcherbina to order an evacuation, the experts were puzzling over how to snuff out the fire in Chernobyl-4. Mayorets mistakenly believed that if workers could pour concrete on the graphite fire inside the reactor and seal it off, then the ambient radiation levels would quickly drop, eliminating the need to evacuate Pripyat. He did not seem to realize that he was dealing with a problem totally outside his field of expertise.

When the panel began brainstorming for a solution to

the reactor fire, everybody was at a loss. The first proposal was to douse the reactor with more water, until somebody correctly remembered that this would be useless. The only way to smother the radioactive fire inside the core was to bury it with sand.

How would they get the sand into the reactor? The commission summoned Major General Nickolai Timofeyevich Antoshkin, deputy commander of the Kiev region's air force, to Pripyat. They instructed him to come at once—and to bring as many helicopters and pilots as he could.

Even as Major General Antoshkin headed for Pripyat, the commission still haggled over whether or not to evacuate the city. The Ministry of Health emphatically urged Shcherbina to relocate people as soon as possible. The ministry's representatives argued, among other things, that most Pripyat residents had not been given potassium iodide, a step that should have been taken immediately. This stable iodine saturates the thyroid gland and prevents it from absorbing the radioactive iodine 131. At this point, the longer people stayed in Pripyat, the greater their risk of developing thyroid cancer. They were also at risk of absorbing high radiation doses from strontium, cesium, and plutonium radioisotopes in the atmosphere. Late in the evening of April 26, Shcherbina finally relented. He ordered the evacuation of Pripyat to take place the next day.

It was too late to coordinate the evacuation first thing in the morning, so the commission would have to settle for an afternoon exodus from Pripyat. During the remainder of the night, some 1,100 buses formed a line along the highway between the town of Chernobyl and Pripyat.

That evening, General Antoshkin arrived in Pripyat, unsure of where to begin. His mission was to drop sand onto the burning reactor via helicopter. In addition to helicopters and pilots, he needed bags and sand. Soon, helicopters started arriving under the cover of darkness. Where would he direct them to land? There was no proper landing field around the Party committee offices in Pripyat. The unflappable General Antoshkin improvised. He climbed to the roof of the Pripyat Hotel and started waving in approaching choppers for landings in front of the Party offices. It became difficult for the commission members to hear each other speak inside the offices, where they conferred into the night. But Antoshkin had no other place to land his growing fleet of helicopters.

Meanwhile, most people in Pripyat still weren't sure just what was going on. So far, only a few details had suggested any real trouble around the nuclear power plant. On the morning of April 26, in addition to the fire at the plant, people had noticed sudsy foam on the street. Nobody had bothered to inform citizens that this foam was intended to help clean dangerous radionuclides off the pavement. One wonders how many children splashed in the radioactive suds as they dawdled on the way to school that Saturday morning. While older children attended classes, younger ones played and rode bikes in the contaminated streets. Pets went outside as they normally would, absorbing radiation through their fur.

That afternoon, the warnings began. Children were advised to go straight home after school. City residents were told to keep apartment windows closed. Not everyone took the warnings seriously, however, because no explanation accompanied them. But plant workers knew something was seriously amiss. When they headed off to start the day shift on April 26, they received an obvious hint: the road leading to the NPP was blocked. Workers coming off the night shift in Chernobyl's other three reactor units had to undergo medical examinations before they could go home. Eventually, the word spread as some people received hushed warnings about the seriousness of the accident from friends and relatives who worked at the plant. These rumors prompted a few people to start leaving Pripyat on their own, and they departed from the doomed town in their own cars, taking radiation with them on the contaminated paneling and tires of their automobiles. But many who stayed were still blissfully unaware that their city was no longer habitable.

By Saturday evening, the shelves of Pripyat's stores were abundantly stocked with red wine. Folk wisdom maintained that drinking red wine would decontaminate people exposed to radiation. In fact, the pre-accident Chernobyl NPP operator's manual recommended it! Some preventative drinking occurred on the streets that Saturday night, where an unspoken tension filled the air. But the apprehension was still unaccompanied by any official explanation of what was wrong at the plant. Nobody told Pripyat's residents that the night of April 26 would be their last in their modern, attractive town that had been custom-built for plant workers and their families. But during the wee hours of Sunday, April 27, residents of Pripyat could see and hear a flurry of helicopters descending on their small city.

When Pripyat residents answered knocks on their apartment doors the next morning, they finally got word of the mandatory evacuation. Buses had pulled up in front of every apartment building in the city by around 2:00 P.M. The residents packed the few essentials they were told to bring. They followed official instructions to bring only enough money and food for three days. A change of clothes was forbidden; clothing was supposed to stay inside closed dresser drawers. As people closed their windows and turned off their

Within a week following the Chernobyl catastrophe, the city of Pripyat was clear of its civilians after officials cordoned off a 30-kilometer area known as the exclusion zone. Only armed guards, sequestered to keep out looters, remained in the abandoned city.

electricity and gas, they also left behind important documents that would have made starting a new life elsewhere a little easier. Because there were official reassurances that they would be returning home within just a few days, the residents did not feel that taking all of their vital records was necessary.

Party workers distributed potassium iodide tablets door to door, although how many people actually got them—or if they received the pills in time to prevent their thyroids from taking in too much radioactive iodine—is still unknown.

Some of the evacuees experienced nausea, headaches, sore throats, and burning eyes that afternoon as they stood outside their homes waiting to get on the buses. Still, they boarded relatively calmly. They reluctantly left their pet dogs, cats, birds, and rabbits behind on the streets. Fur and feathers absorb radiation easily. Therefore, pets, especially those that went outdoors daily, were too contaminated to leave Pripyat with their owners. Grigory

Medvedev described what happened after the buses pulled away:

> Some dogs, finding themselves left behind by their masters, ran after the buses for a long way, but to no avail. Eventually they fell back and returned to the abandoned town, where they began to roam around in packs. First they devoured a large number of the radioactive cats and then turned wild and began to growl at humans. They attacked humans and abandoned farm animals a number of times.

From April 27–29, volunteer hunters shot down the abandoned, radioactive dogs prowling Pripyat's empty streets. Ironically, the dogs that were killed quickly were the lucky ones: animals who evaded the shooters developed radiation poisoning, went blind, and died painfully within a month or so.

The animals' former owners struggled to find places to stay. The caravan of buses stayed together for some 37 miles. When they reached the city of Ivankov, however, the buses scattered in search of temporary homes for their passengers. Some evacuees got off the buses and started walking. Nuclear refugees plodded along the highway towards Kiev, about 160 km south of Chernobyl, perhaps hoping to stay with friends or relatives in the Ukrainian capital. The evacuees had only the clothes on their backs and a few days' worth of cash on which to get by.

It was apparent within days that nobody would be returning to Pripyat for a long time—and that many more people still faced mandatory relocation. By May 2, a 10-kilometer radius around the power plant was bereft of people; by May 5, a 30-kilometer area, called the exclusion zone, was emptied of all humanity except armed guards ordered to keep out looters. At least, that is what everyone thought until two elderly women were discovered in their

Pripyat apartments over a month after they were supposed to have left with the other evacuees. The women were admitted to the hospital after their month in hiding.

Frantic efforts to contain the fire raging in the gutted belly of Chernobyl-4 had continued to no avail for days after the evacuees rolled out of Pripyat. General Antoshkin spent much of April 27 securing sand for his mission even as helicopters continued to land outside Pripyat's Party offices. Plant operators filled bags with sand from the grounds around the plant. The job was extremely hazardous, since radiation concentrates heavily in sand particles. Some farmers working in the surrounding area were forced to stop their own jobs and help fill bags with sand. While this makeshift crew of workers, called *liquidators,* scooped up radioactive sand, helicopter pilots hovered above Chernobyl-4, trying to pinpoint their target. It was actually very difficult for anyone unfamiliar with the reactor unit in its pre-accident state to find the reactor core while looking down on the wreckage from a chopper. The pilots needed operators who knew Chernobyl inside and out to accompany them on their "bombing" missions, during which they would drop a mixture of sand, lead, and a mineral called dolomite onto the burning reactor core.

The pilots approached the reactor at an altitude of about 110 meters. Once they were over the burning mouth of Chernobyl-4, they had to hover directly above it long enough to drop bagfuls of the sand mixture. The radiation level in the air above the reactor was approximately 500 roentgens per hour during those first days after the explosions. After just one or two sorties to the burning reactor unit, the pilots and their navigators would become too sick to fly anymore. The pilots tried to limit their radiation exposure by sitting on lead plating as they aimed for the "red glow"

Helicopter pilots had a similar view of the Chernobyl-4 reactor shaft when they dropped large amounts of a sand mixture onto the graphite fire in the reactor's core. The pilots had only a few moments to complete their task before exposing themselves to deadly radiation.

inside the core and dropped the sand mixture.

Helicopter pilots flew more than 1,800 missions to the wrecked reactor building between April 27 and May 9 as the fire continued to spew radioactive gases into the air. They dropped some 5,000 metric tons of sand, dolomite, boron carbide, and other neutron suppressors and absorbers on the stubborn inferno. Victory over the nuclear fire inside Chernobyl-4 seemed close at hand on May 9, when radioactive emissions from the reactor unit began to decrease. Everyone monitoring the site breathed a sigh of relief; perhaps the inside of Chernobyl-4 had finally stopped burning. Their relief was short-lived, however. A small spot inside the reactor's wreckage still glowed bright red. Valery Legasov wrote that 80 tons of lead were deposited into the crater to extinguish the spot. The fire finally went out on May 10.

In hindsight, it is doubtful whether the helicopter missions helped to extinguish the fire at all. The "red glow" the pilots had been instructed to hit was actually about 50 feet away from the core, according to Alexander Sich, an American nuclear engineering student from the Massachusetts Institute of Technology who conducted the first in-depth study of what really happened after the Chernobyl disaster. Sich, who is of Ukrainian ancestry, spent about 18 months around the accident site during the early 1990s conducting extensive research for his doctoral thesis. He concluded that the core was never smothered but instead melted down completely on its own. The liquefied core material spilled out of the reactor and down onto the concrete floor of the reactor room, where it spread out thinly enough to arrest the fission reaction. This allowed the fire in the core to eventually just burn itself out.

The end of the fire did not mean that the dangers plaguing the mangled reactor unit had ended, however. It buckled under the weight of the sand and lead poured into it. The liquidators' hazardous work had just begun.

The tasks the liquidators tackled were numerous. Their level of risk varied with the types of duties they performed. Shoring up the reactor unit, now filled with tons of sand and lead—with liquefied core material underneath it—was an extremely hazardous assignment. Perhaps the most dangerous task, however, was removing the intensely irradiated graphite fragments

In the summer of 1986, workers pose in front of the uncompleted "Sarcophagus," a structure constructed to seal up Chernobyl-4. The banner written in Russian reads, "We will fulfill the government's order!"

that the explosion had scattered all over the roof of Chernobyl-4's turbine hall. Attempts to manipulate robots to pick up the graphite had failed. The robots became ensnared in hoses left behind on the roof by the firefighters. They also proved impossible to operate in fields of high ambient radiation. Human liquidators—eerily termed "bio-robots"—wearing heavy protective clothing and respirators had to pick up the graphite by hand. The bio-robots' radiation exposure was so high

This aerial view of Chernobyl-4 shows the ravages of the explosion on April 26, 1986. However, as cleanup workers and citizens in surrounding areas would soon learn, the real danger of the disaster came in the form of exposure to deadly radiation.

that even in protective gear they could only work in shifts lasting approximately 90 seconds.

Some liquidators assisted health care workers. Others washed streets and buildings in an effort to decontaminate them. Liquidators deforested the woods around Pripyat because foliage absorbs radioactivity extraordinarily well. They buried the radioactive trees and foliage in pits. They removed contaminated topsoil in outlying areas and replaced it with "clean" soil to save some farmland. One of the liquidators' most risky and important endeavors was the construction of the "Sarcophagus," a concrete and metal structure meant to seal off Chernobyl-4 for about 30 years.

The liquidators with the most dangerous jobs were paid handsomely with cash on the spot. Although many of them joined the cleanup for the money, others were conscripts from the Soviet army. For them, the grim alternative to cleanup work at Chernobyl was fighting in Afghanistan, where the Soviets were trying to put down anti-Socialist rebels.

Whether motivated by money or bound by duty, however, many liquidators possessed little understanding of the risks they were taking. Some liquidators poked holes into their respirator masks so that they could smoke cigarettes through them. Others left their masks dangling around their necks, oblivious to the fact that they were inhaling radionuclides with each breath. In *The Truth about Chernobyl,* Medvedev recalls an especially jarring example of ignorance: he encountered a young woman in a liquidator's uniform picking blossoms and sniffing them deeply. When Medvedev warned her that the flowers were contaminated, she responded, "Come on!" and proceeded to pick some more.

Whether some people working around the stricken NPP flouted safety procedures out of ignorance or out of skepticism for the half-hearted warnings authorities gave them about radiation safety is a good question. After all, they were residents of a country whose leaders did not bother to tell neighboring nations what had happened until a radioactive cloud from Chernobyl-4 made hiding the awful truth impossible.

Although the USSR took measures to protect its citizens by evacuating them from the area around Chernobyl, long-term contamination and dangerous levels of radiation in the environment continue to plague the area to this day. Workers examined this woman's tomatoes from the Ukraine, which exceeded normal levels of radiation by 20-30 times.

# Cloud of Secrecy

By the morning of April 28, evacuation efforts around the Chernobyl NPP were well under way. Although the USSR was finally taking measures to protect those who lived around the Chernobyl NPP from the undeniably dangerous levels of radiation there, the Soviet government was not alerting its neighbors to the disaster. This lie of omission first started to unravel at Sweden's Forsmark nuclear power plant.

As workers entered Forsmark to start the day shift on Monday, April 28, excessive radioactive contamination was detected on their clothing, shoes, and hair. Once plant directors were certain that there were no radiation leaks inside Forsmark, they concluded that the radioactivity was coming from outside the plant. Swedish authorities soon suspected that the extreme radioactivity was coming from their neighbor to the south, the USSR.

When Sweden asked officials in Moscow if the Soviet Union was the source of the sudden increase in radioactivity, their query was met with categorical denial. But United States intelligence efforts unintentionally exposed the cover-up. A U.S. spy satellite surveying Soviet nuclear missile silos had orbited over the Chernobyl region just 28 seconds after the explosions on April 26. When the U.S. government saw the satellite image of the accident—which showed an explosion, but no missile launching—it assumed that a missile had accidentally detonated on the launchpad. It wasn't until after checking a map to pinpoint the explosion's location that the United States realized that the satellite image was not of a missile detonation at all: disaster had struck at the Chernobyl NPP, and it was visible all the way from outer space.

The evidence of trouble at Chernobyl was irrefutable, yet the Soviet government remained silent for hours after Sweden's inquiry. Finally on the night of April 28, TV news in Moscow carried this brief and largely uninformative statement: "An accident has occurred at the Chernobyl Atomic Power Plant and one of the reactors was damaged. Measures are being taken to eliminate the consequences of the accident. Those affected by it are being given assistance. A government commission has been set up."

In a second statement issued on April 28, the Soviets attempted to distract the world community from the serious disaster at Chernobyl by discussing the shortcomings of the United States nuclear power industry—with special mention of the 1979 accident at Three Mile Island. Yet the next day, the Soviet Union sought expert advice from both West Germany and Sweden on extinguishing a graphite fire.

In the absence of detailed, verified information from the USSR about the accident, the world media filled in the blanks. False stories of mass graves in the woodlands surrounding Chernobyl appeared in newspapers around the globe. Some unscrupulous individuals exploited the tragedy by selling doctored photos and spurious film footage to news agencies. Still, the Soviet Union did not come forward to set the record straight.

As other nations discovered that they, too, were affected by the Chernobyl accident, the international community grew increasingly angry. The disaster's impact extended beyond Soviet borders, yet the USSR behaved as if the accident were merely an embarrassing slip-up to be concealed from the world. This decision to hide the severity of the accident sparked a worldwide furor.

The explosion of Chernobyl-4 released irradiated particles of core material, such as fuel assemblies and graphite moderator blocks, and roofing into the air. Many of these hot particles fell in the vicinity of the plant. But the rest rose an estimated three kilometers into the air and formed a deadly radioactive cloud. The wind swept up this plume and sent it on a journey that lasted over a week. The hot particles rode on a prevailing northwesterly wind to Latvia, Poland, and Scandinavia. The cloud then spread out, going west and southwest to Austria and parts of Germany, Italy, and France. In five days, the plume was heading northwest again, reaching the middle of Germany, Great Britain, and the Netherlands. In about a week, it was starting to dissipate as it moved northeast. Even Canada and the United States received tiny amounts of Chernobyl fallout. In all, nearly two dozen countries were affected by the radioactive plume.

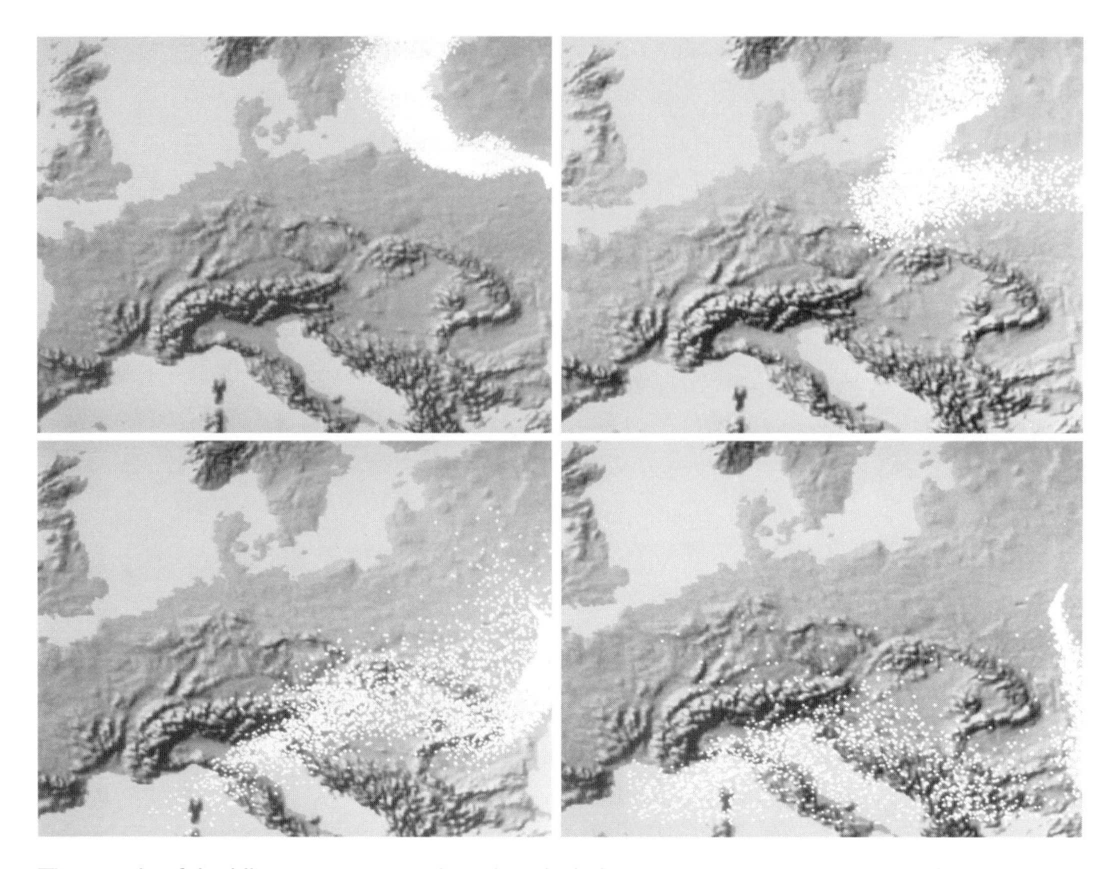

The people of the Ukraine were not the only ones affected by the Chernobyl disaster. The diagram above tracks the path of the radioactive cloud that drifted from the Ukraine, across the USSR, and through several European countries. In the years following the catastrophe, many medical experts would make connections between increasing cancer rates and the radioactive material from this cloud.

The cloud didn't spare Soviet cities, of course. In Kiev, May Day celebrations went on as planned on May 1. According to physician, writer, and Ukrainian ambassador to the United States Yuri Shcherbak, radiation in that city was about 100 times higher than safe levels when people attended the outdoor festivities in lightweight clothes and short sleeves. In *The Legacy of Chernobyl*, radiobiologist Zhores Medvedev wrote that Kiev should have been evacuated by April 29 or 30 for maximum safety.

It wasn't until May 15 that Kiev began evacuating children under age 14, mothers of small children, and pregnant women. The children and the mothers of the littlest ones crowded into Young Pioneer summer camps outside of the city, where they learned what they could about the accident via television. City schools closed for the summer on May 14 to accommodate the evacuation. The children of Party officials, however, had begun disappearing from Kiev classrooms days before the official evacuation. In the weeks after the accident, 31 people died in Kiev from radiation poisoning.

The radioactive plume's fallout was spotty because the amount of contamination a given area experienced depended on weather and wind conditions during the time that the plume passed overhead. If it rained where the plume was, for example, the affected area might become a hot spot, a pocket of heavy contamination. Had it been raining in the Ukrainian capital of Kiev when the plume came, the number of fatalities in the weeks after the accident probably would have been much higher than 31.

As European nations nervously watched the progress of the radioactive plume, people were advised to stay indoors with their windows closed, to shower frequently, and to wash produce thoroughly before eating it. Poland tried to safeguard the health of children up to 16 years old by making doses of stable iodine available to them to prevent thyroid cancer. In the United States, the Environmental Protection Agency (EPA) increased its checks for air pollution in particle form, monitoring the air daily instead of the usual two times per week in case significant fallout from Chernobyl reached America.

Even as the radioactive plume from Chernobyl terrified people outside Soviet borders, the USSR kept quiet. Secrecy was a firmly entrenched habit in the Soviet Union. Although Mikhail Gorbachev, general secretary of the Communist Party, had pledged a new era of *glasnost*, or openness, in Soviet newspapers, TV, and radio, he maintained a low profile regarding Chernobyl. Although Soviet reports never hesitated to seize on accidents at Western nuclear power plants, the Communist Party had historically hushed up similar problems in the USSR. Numerous accidents had occurred prior to Chernobyl, but none so far-reaching and difficult to conceal. For instance, 14 operators were boiled to death less than one year before Chernobyl at the Balakovo NPP in Russia during the start-up of its Reactor One. This accident went largely unreported.

Just weeks before the Balakovo accident, Minister Mayorets had expressly forbidden the publishing or broadcasting of any news about "the unfavorable ecological impact of energy-related facilities (the effect

## OTHER NUCLEAR ACCIDENTS

America's most serious commercial nuclear power plant accident occurred on March 28, 1979, at the Three Mile Island (TMI) facility, just outside of Harrisburg, Pennsylvania. At 4:00 A.M. water stopped flowing to the steam generators in TMI-2, a reactor cooled and moderated by pressurized water. An operator turned on emergency feedwater pumps to cool the overheating reactor core, but he was not aware that very little water was actually reaching the core because valves to the emergency pumps were stuck in the closed position. Two indicator lights went on in the control room to alert operators to the malfunctioning valves. Nobody noticed these warning lights, however, because a maintenance tag was covering them.

The safety systems in TMI-2 shut down the reactor when pressure in the core rose to dangerous levels. But this occurred after a pressure release valve at the top of the reactor had already opened. Once pressure in the core decreased, this release valve failed to close completely. The reactor's emergency core cooling system fed water into TMI-2, but the operators disabled the cooling system when they thought the danger was past. The core overheated, causing much of the fuel in TMI-2 to melt. The operators did not realize that all the water they had fed to the reactor was leaking through the partially opened pressure release valve and onto the floor of the containment building. Radioactive gases from the melted fuel contaminated the leaked water and escaped into the environment.

of electromagnetic fields, irradiation, contamination of air, water, and soil) on operational personnel, the population, and the environment."

Although the disaster at Chernobyl proved impossible for the Soviet Union to cover up, the government still tried to downplay its seriousness. A press conference held on May 6 by the Soviet Ministry of Foreign Affairs typified this effort. The graphite fire was still burning in the reactor at that time. Nobody at the press conference was willing to clarify the condition of Chernobyl-4, however. Soviet and Eastern European journalists were permitted to ask spoken questions, but Western journalists

Although radiation doses to people near TMI were estimated to be minimal, the news did little to calm those residents who reported nausea, a metallic taste in the air, and the deaths of pets and wild birds after the accident. Pregnant women and young children living very close to the reactor were advised to temporarily evacuate the area—and many others fled of their own accord. The operator of TMI at the time of the accident, GPU, Inc., permanently shut down TMI-2. The company has reportedly paid residents nearly $4 million in settlements so far.

Nuclear accidents are not confined to power plants. Three workers in Tokaimura, Japan, were mixing uranium with nitric acid at a fuel fabrication plant on September 30, 1999. To prevent the uranium from forming a critical mass and starting a chain reaction, the workers were supposed to use narrow pipes to pour it into the tank. They used buckets instead. The resulting fission reaction eventually killed one of the three workers, seriously injured another, and imperiled the last worker's future health. Dozens of other plant workers and several emergency response personnel were also irradiated.

Five hours after the accident, Tokaimura authorities ordered 160 people who lived near the plant to evacuate their homes; they advised those within 10 km of the facility to stay inside. The advisory was in effect until October 1, although those closest to the plant were kept away longer. The Japanese government's slow reaction to the crisis provoked international criticism.

could only submit theirs in writing beforehand. The lack of clear answers regarding the status of Chernobyl-4 left even journalists from "friendly" nations uncertain of what was happening at Chernobyl.

Details on the accident remained sketchy until May 14, when Mikhail Gorbachev appeared on Soviet television to finally admit that the accident was indeed gravely serious. He said that nine people had died since April 26. "Extensive work still lies ahead," Gorbachev said of the efforts to liquidate the disaster's consequences. But the general secretary of the Communist Party also used his TV appearance to take a swipe at the "most dishonest and malicious lies" posited by the rest of the world about conditions at Chernobyl in the absence of official Soviet information. In addition, Gorbachev tried to get U.S. president Ronald Reagan to agree to a test ban on nuclear weapons in the wake of the accident. The cloud of secrecy around Chernobyl was slowly dissipating, but not without exposing some of the USSR's old grudges against the West.

Was the Soviet Union's attempt at candor about Chernobyl too little, too late? The health—perhaps even the lives—of many local people might have been saved by honesty from the start about the invisible, odorless danger wafting out of Chernobyl-4. As it was, 36 precious hours elapsed between the destruction of the reactor and the evacuation of Pripyat. Kiev was not evacuated for weeks.

By June, the Party found itself facing a backlash of criticism from a most unlikely source: the Soviet media. Print, radio, and TV in the USSR were controlled by the Communist Party. This meant that criticism of the government practically never made it into print or onto the air. The media traditionally told of happy, productive Communist workers and safe, efficient farms and

As general secretary of the Communist Party, Mikhail Gorbachev was the first leader to appear on television and address the Soviet public about the magnitude of the Chernobyl disaster. In anticipation of foreign criticism of the USSR's nuclear safety standards, Gorbachev reminded his audience that the West had had its own nuclear power debacles.

industries. Under General Secretary Mikhail Gorbachev, however, articles and newscasts that expressed doubts about the government's handling of Chernobyl suddenly started to emerge in the USSR. Television news crews visited the NPP and interviewed liquidators and evacuees without fear of reprisal by the Party.

The USSR prepared a 430-page report on the causes and consequences of the Chernobyl accident for a meeting on the subject by the International Atomic Energy Agency (IAEA). This meeting, which took place in Vienna, Austria, in August 1986, included delegates from 62 countries. Many were surprised by the degree of openness they discerned in the Soviet report. Still, the Soviet delegation of the IAEA, led by Valery Legasov, doggedly defended the RBMK reactor design. Legasov made a lengthy presentation at the convention. He placed blame for the accident squarely on the shoulders of the operators at Chernobyl. "The defect of the system," said Legasov, "was that the designers [of the RBMK] did not foresee the awkward and silly actions by the operators."

The case of Valery Legasov goes to show that not everyone in the USSR could afford to be candid during the first flowering of glasnost. Legasov was in line to succeed the aging director of the prestigious Kurchatov Institute of Atomic Energy, Anatoly Alexandrov, who had designed the RBMK nuclear reactor. Legasov risked derailing his career if he blamed the accident on faulty reactor design.

Although the Soviet delegation's report to the IAEA was more revealing than many delegates from the 61 other participating nations expected it to be, it still bore the stamp of old-style Party tactics. It was clear in the report's conclusion that less-powerful members of the Communist Party would be scapegoats to shield high-ranking officials from blame, and that the operators alone would be responsible for the accident.

In July 1987, the abandoned town of Chernobyl's Dom Kulturi (Culture House) reopened briefly for a high-profile courtroom trial. The six defendants included Viktor Petrovich Bryukhanov, Nickolai Masimovich Fomin, and Anatoly Stepanovich Dyatlov. The only (low-level) government official on trial was an inspector named Yuri Laushkin, who stood accused of failing to ensure that the Chernobyl NPP complied with established safety regulations. No high-ranking Party officials faced charges related to Chernobyl. While it is true that Bryukhanov, Fomin, and Dyatlov did share responsibility for the deaths at Chernobyl, many of their actions resulted from fear of punishment by their superiors in the

Party. They were charged with multiple safety violations.

The officials also faced charges for waiting 36 hours before notifying the public of the dangers posed by the accident. One charge against Dyatlov stemmed from his refusal to believe that Chernobyl-4 had been destroyed; he was held responsible for the deaths of four workers he sent to check on the reactor. The trial's opening arguments were televised in the USSR. When it was over, Bryukhanov was sentenced to 10 years in a labor camp, although he was released in 1991. Fomin also got a 10-year sentence, but he was released early for psychiatric problems. He even went on to work in another nuclear power plant! Dyatlov was sentenced to 10 years of labor camp, but just like Bryukhanov and Fomin, he got out early, for declining health. Yuri Laushkin, the government inspector, got a shorter sentence, was released early, and died of stomach cancer two years later. The two other defendants were NPP technicians who also got shorter sentences than Bryukhanov, Fomin, and Dyatlov. Many more Chernobyl operators were dismissed or demoted, and some of them were also expelled from the Communist Party.

*Time* magazine characterized the trial at Chernobyl as "a hardheaded exercise in damage control." The trial, like the IAEA report nearly one year earlier, in an attempt to salvage the reputation of the Soviet nuclear energy program as a whole, blamed lower-level individuals—who were certainly liable, but not solely responsible—for the accident.

# A Far-Reaching Disaster

A sign posted in the Polissya State Radioactive/Ecologic Reserve, located on the Belarusian side of Pripyat, warns that the area is radioactive. Only scientists and a few workers are allowed in this section of the forest, which remains the only radioactive nature preserve in the world.

# 5

Throughout the summer of 1986, the evacuations continued with no end in sight in the areas most heavily contaminated. After Pripyat and Chernobyl were abandoned, the rest of the exclusion zone was emptied of residents. By September 1986, approximately 116,000 evacuees had permanently left their old homes in the Polissya of Ukraine, Belarus, and Russia. The Polissya is a lush region of swampy forests and prime farmland that for centuries had been the breadbasket of the Soviet Union. Many of these people wound up being relocated to Kiev. There, they faced new neighbors who were sometimes less than sympathetic to their plight. Evacuees received taunts of "glow worm" or "firefly," and children were ostracized by their new schoolmates. The Soviet government hurriedly built new housing to accommodate the refugees. By the end of 1987, approximately 21,000 new houses in Ukraine, 15,000 of them in Kiev, were allocated to evacuees. Some Kievans watched resentfully as the Ukrainian government allotted brand-new apartments to Chernobyl evacuees who now urgently needed them—after they themselves had been forced to languish for years on waiting lists for new housing.

The abandoned 30-kilometer exclusion zone, or alienation zone, is not as chillingly empty as its name implies. Wild boars, otters, bison, deer, fox, and once-endangered gray wolves now thrive in the Polissya State Radioactive/Ecologic Reserve, located on the Belarusian side of the zone. The world's only radioactive nature preserve, Polissya is closed to everyone but scientists and a handful of workers. The Chernobyl accident irradiated some 1,500 acres of forest around Pripyat, forcing liquidators to uproot the trees and bury them in clay-lined pits. Inside the Polissya Reserve, however, the vegetation is lush and the wildlife abundant.

Unfortunately, the vegetation and wildlife are also radioactive, as studies of the animals in the preserve have shown. The animals' radiation exposure is undoubtedly changing their genetic makeup. In 1995, a group of researchers from the United States found that field mice inside the exclusion zone displayed a surprising degree of genetic mutation. In fact, their genes more closely resembled those of rats than they did the genes of other field mice living outside of the exclusion zone. Mice and rats have been two distinct species for many millions of years. Yet the mutations that had occurred in the Chernobyl mice within just one decade of the accident had slightly blurred that ancient distinction. Such mutations aren't limited to rodents. When Swedish scientists compared swallows nesting in the radioactive reserve to swallows in less irradiated areas 100 km away, they observed a much higher incidence of albinism (a hereditary lack of pigment in the feathers, skin, and eyes) in the birds living inside the exclusion zone.

Many researchers who visit the zone seek to better understand the effects of chronic radiation exposure on humans by studying the wildlife at Polissya. With less human interference in the exclusion zone, wild game animals now have a chance to make a comeback in an area once dotted by small towns and villages. Eating any of the game animals in the preserve, however, would be foolhardy because their meat would be dangerously radioactive.

The plight of the Saami (or Lapp) people of northern Sweden illustrates how the Chernobyl accident damaged the entire food chain in affected areas. Reindeer herding is a central part of traditional Saami culture. When Chernobyl's radioactive plume was overhead in Sweden, spring rains deposited excessive amounts of cesium 137 in the soil. Lichen, a type of moss favored by the Saami's reindeer herds, became highly contaminated. The reindeer feasted on the irradiated lichen, and their meat was soon judged unsafe to eat or to sell.

The contamination of tons of reindeer meat was both an economic and cultural disaster for the Saami. In addition to relying on the meat for their own subsistence, they also sold it. The Swedish government raised the legal limit on radioactive contamination levels in reindeer meat for the Saami's consumption, but bans on the sale of the meat devastated their livelihood. By the spring of 1987, some 50,000 reindeer had been slaughtered and buried in pits because they were too contaminated to eat even under the government's more liberal standards. Some irradiated reindeer meat was eventually sold to fur farms as animal feed. Since animals at fur farms spend their entire lives in captivity and are not raised to feed people or other animals, the contaminated meat they ate stayed out of the

European wildlife, in addition to people, suffered from Chernobyl fallout. The habitat of northern Sweden was hit hard after thousands of reindeer ate irradiated lichen. With reindeer meat as one of its primary products, the Swedish meat industry also faced setbacks: by the spring of 1987, it deemed some 50,000 reindeer to be contaminated.

natural food chain. But sales of reindeer meat for human consumption were banned for years in heavily contaminated areas. Once the heart of Saami life, the semi-nomadic existence of the herders is threatened today—and Chernobyl is partly to blame.

Like the reindeer of Swedish Lapland, human beings risked ingesting harmful levels of radiation if they ate fruits, vegetables, or grain products cultivated in contaminated areas. Over 20 countries received radioactive fallout from Chernobyl. Restrictions on foodstuffs ranged from reminders to wash all fruits and vegetables to complete bans on the importation and sale of leafy vegetables grown in affected areas. These restrictions mainly affected Soviet farmers, but some farmers in Italy, France, and Greece also suffered. In the early post-accident days, the sight of a dosimetrist checking produce became common at many European food markets.

When farmers within Chernobyl's 30-kilometer exclusion zone were evacuated, most of their livestock was shot. Many of the farmers were incredulous, since their cows, sheep, and goats looked perfectly healthy. But some 86,000 head of cattle were evacuated from the exclusion zone to "clean" pastures. If they grazed on clean grass for several months, their meat could return to safer radiation levels as they metabolized the radioactive particles they had ingested.

Cleanup workers disposed of the topmost layers of agricultural soil in an effort to salvage some land for commercial crops. Still, many farmers would eventually leave, unable to sell anything they grew. Others would keep growing subsistence crops for their own families. The Polissya was largely rural with isolated villages. Prohibitions against growing crops on small family plots were hard to enforce under these conditions. In addition, many rural people still rounded out their diets with wild mushrooms, berries, and fish—more foodstuffs contaminated by the accident.

Even if people avoided eating fruits and vegetables grown in contaminated areas, they would still ingest radionuclides if they consumed meat from animals that grazed in contaminated pastures or if they drank milk from cows irradiated in this way. Although iodine 131 has a *half-life* of just eight days, people were worried right after the accident since the isotope not only contaminates soil, but it also enters cow's milk via any contaminated grass the cows eat. The Soviet government quickly

Despite the Soviet government's attempts to curtail the long-term effects of Chernobyl fallout, they could not completely control all of its citizens, as is the case with this dairy farmer and resident of a radioactive zone, drinking fresh milk from what may be a contaminated cow.

enacted restrictions on milk production and consumption. In some affected areas of the Soviet Union, only state-run stores could sell milk after the accident. But many farmers who either lacked access to a state-run store or were too poor to buy milk at government prices continued to use milk from their own cows.

The devastation Chernobyl inflicted on the Soviet Union's livestock was not confined to the early post-accident weeks. Farmers in Ukraine and Belarus reported a drastic increase in the number of deformed piglets, calves, and foals born to their animals within less than two years of the accident. Among the problems suffered by affected animals were extra or missing limbs and missing eyes, ribs, hair, and anal openings. The rise in animal birth defects near Chernobyl terrified people still living in contaminated areas.

Evacuations continued long after the initial exodus of

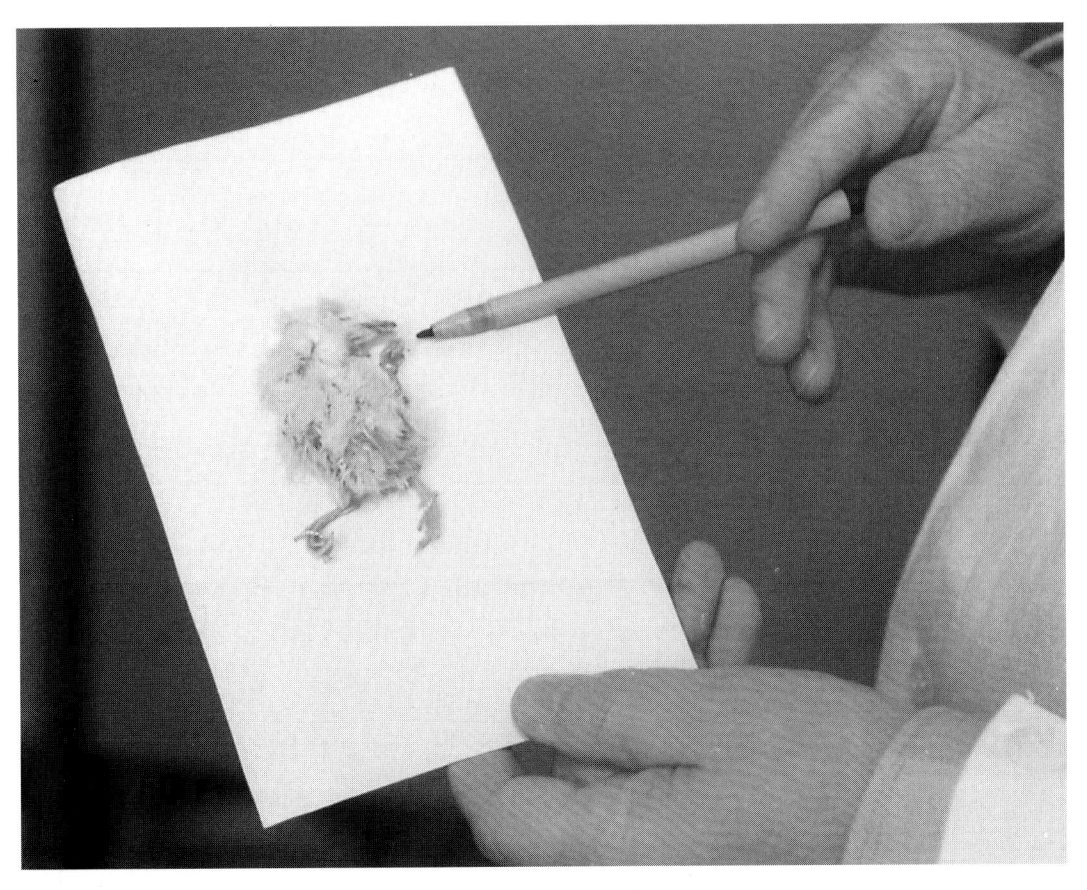

A scientist from the Ukrainian Academy of Sciences demonstrates with a four-legged chick the biological effects of radioactivity. After the Chernobyl catastrophe, many researchers discovered genetic mutations, as well as new bacteria and viruses, in the farm animals of exposed regions.

about 116,000 people was completed in September 1986. In 1988, with no end to planned evacuations in sight, authorities tried to stem the tide of environmental refugees by raising the maximum lifetime dose limit for people not working in the nuclear industry from 25 to 35 rems. This measure did little to slow down the flow of evacuees, though. It didn't take long for radiation and medical experts—who by now scrupulously monitored the physical condition of people living around the NPP—to notice that many people still awaiting evacuation were fast approaching the revised dose limit of 35 rems. They would have to leave their towns and settlements soon.

Richard Mould, author of *Chernobyl Record*, wrote that 53,000 Ukrainians, 107,000 Belarusians, and 50,000

Russians were evacuated from 1990 to 1995. He estimated that by 1998 at least 284,000 people had been removed from their homes because of radioactive contamination. A small proportion of the early evacuees, however, were eventually allowed to return to their original towns and villages, thanks to cleanup efforts.

Many evacuees have suffered emotionally and socially since leaving their homes. Some people were merely sent from one side of their district to a less contaminated place on the other side. Others, however, were relocated to Siberia or the Balkans. Evacuee families had no control over where they ended up. This compounded a sense of helplessness among many evacuees, who were already worried about the future effects of radiation exposure on themselves and their children.

In 1992, older people without children or grandchildren at home were legally permitted to move back to their abandoned houses in some parts of the exclusion zone that were at least 10 km from the NPP. An estimated 600–700 persons were living in the exclusion zone for at least part of the year by 1998.

People had been sneaking back into the exclusion zone long before they got permission to do so. The Ukrainian side of the zone is fenced and patrolled by guards, but the area just outside Pripyat is wooded, permitting people with sufficient determination and familiarity with the forest to sneak back home. Some senior citizens who had learned to navigate the woods during World War II did just that. The fall of Communism in 1991 and the subsequent breakup of the Soviet Union into independent republics made returning to the exclusion zone much easier, since security at the boundary slackened considerably.

Most people living inside the exclusion zone have chosen to be there. Others still must live and/or work in areas contaminated enough to necessitate continued dietary restrictions and dosimetric monitoring. These areas, known as Strict Control Zones (SCZs), are parts of Ukraine 200 km or more from the Chernobyl NPP that are still deemed habitable despite high radioactivity. There were 412 SCZs in 1988. Although the people living in SCZs did not have to leave their homes, their food was monitored for radiation and they received financial compensation for the inconveniences and risks that living in a contaminated zone entailed. People living in Strict Control Zones collected regular payments of 30 rubles per family member for the purchase of "clean" food at state-run stores. Some workers who conducted business in the SCZs received benefits that virtually doubled their pre-accident salaries.

But the people who received the food allowances and benefits had a scathing term for these payments: "coffin bonuses." Many people viewed the compensation as evidence of the government's knowledge that they were doomed if they remained in the SCZs. The benefits also generated bad feelings between certain individuals. The uneven distribution of radioactive fallout from Chernobyl meant that in some cases one family was eligible for SCZ benefits while their neighbors were not. Many people judged ineligible for SCZ benefits were bewildered and angry because people living or working practically next door to them collected benefits while they did not. Thus, the payments—intended to compensate people in SCZs for the daily inconvenience of following laws about what they ate and for putting up with continued dosimetric monitoring—actually stirred up increased feelings of doom and helplessness in those who got them. The irregular, seemingly arbitrary boundaries of the SCZs also angered people who felt they were forgotten by the Soviet government.

Fearing for their health and tired of government supervision, many professionals who could afford to move left the SCZs. Farmers left, too, frustrated by their inability to make a living with food restrictions in place. Other districts and towns outside the SCZs also had restrictions on locally grown food that lasted into the 1990s.

After the first deaths from radiation sickness and the evacuation of the exclusion zone, the true magnitude of the accident's damage to the environment began to reveal itself. In the Polissya region of Ukraine, Belarus, and Russia, life had changed relatively little for hundreds of years in many of its small farming villages. After the destruction of Chernobyl-4, however, many family farmers found their time-honored way of life destroyed. They became nuclear refugees, shunted into city apartments with no community to support them.

Much of their once-abundant land is useless. The closure of the exclusion zone permanently took with it about 70,000 hectares of agricultural land. (One hectare is slightly less than 2.5 acres.) Between 100,000 and 150,000 additional hectares of farmland outside the exclusion zone are too contaminated for cultivation—and probably will be for centuries to come. Liquidators tried to decontaminate about 2 million hectares of land by disposing of contaminated soil and replacing it with clean earth.

It is uncertain whether these attempts to contain the disaster will be effective in the long run. Although drainage barriers were built to prevent water from the NPP cooling pond from entering the Pripyat River, there are some 800 waste disposal sites in the exclusion zone that contain radioactive cars, buses, bulldozers, helicopters, and other debris from the cleanup efforts. Many of these sites were little more than holes dug hastily into

The Soviet government put these Soviet military and emergency vehicles and hundreds more out of commission after they were irradiated in the Chernobyl cleanup.

the ground without protective concrete linings: some of these now hold vast pools of radioactive water. How have these haphazard waste sites affected the underground water table? There is no clear answer to date.

Even worse, much remains unknown about the effects of radiation exposure on human health. The people affected by Chernobyl's deadly cloud faced years of anxious waiting and watching over their own health and that of their children. A wait-and-see approach would be necessary, since the blowout of Chernobyl-4 had no precedent. How many people would get sick as a result of the accident? How many would die?

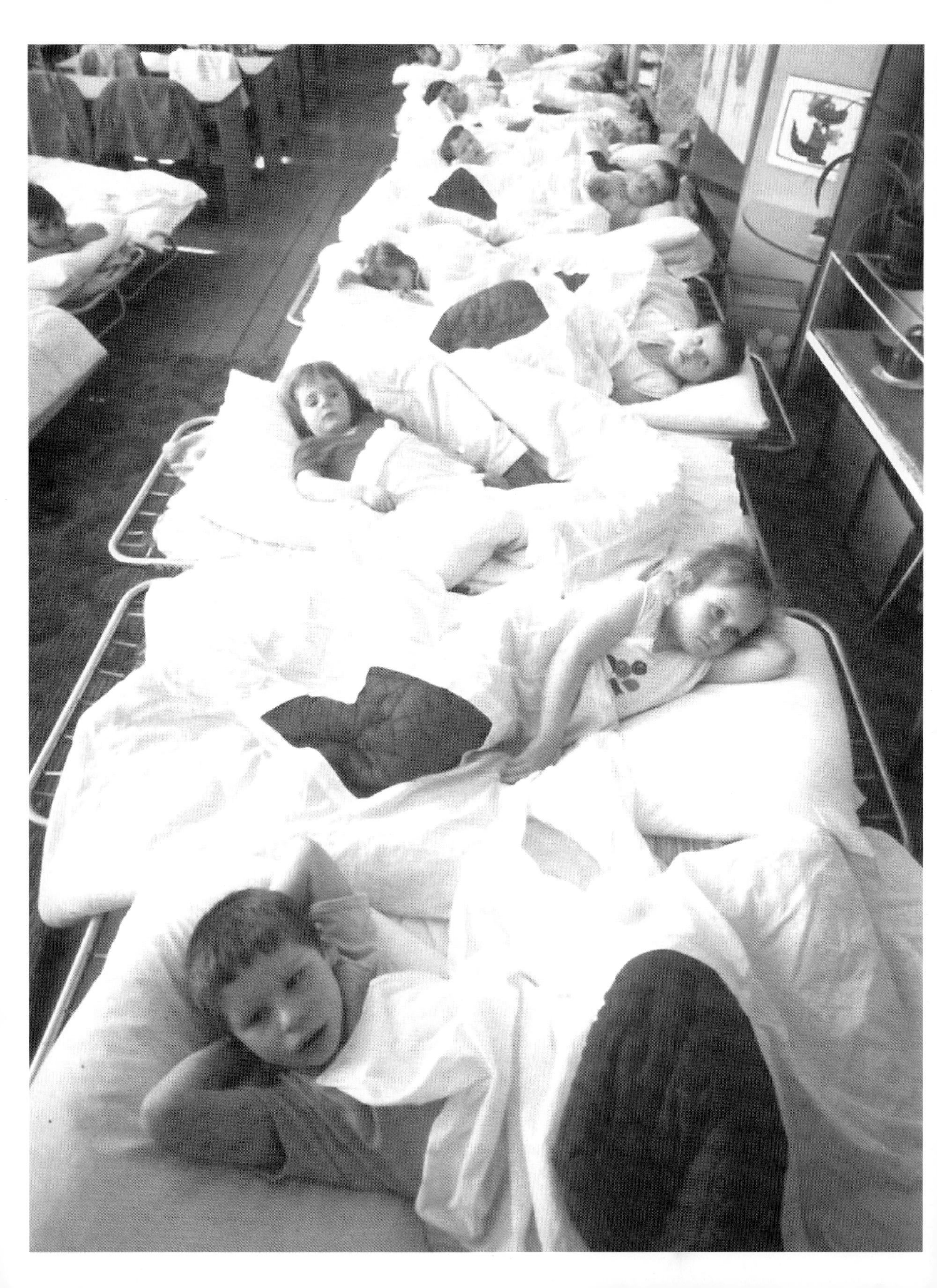

Four years after the 1986 accident, these young patients of a hospital in Syekovo, a village near the Chernobyl plant, were still suffering from radiation-related illnesses. At the time the children of this ward were receiving treatment for intestinal problems.

# 6 Health Effects of Chernobyl

The disaster at the Chernobyl NPP turned the surrounding area into a laboratory for the grimmest of experiments: a long-term study of radiation's effects on the human body. How do people absorb radiation? There are five basic paths: external exposure (being outdoors when a radioactive plume is overhead, for example); inhalation of radioactive particles in the air; beta radiation burns to the skin; external contact with radioactive material deposited on the ground (radioactive dust on shoes and clothing, for instance); and ingestion of contaminated food or beverages. People absorbed radiation via all five paths after the Chernobyl accident.

A vast array of radionuclides escaped from the ruptured top of Chernobyl-4, all of which emitted radiation, in the form of alpha and beta particles and gamma rays. Although alpha particles have virtually no ability to penetrate flesh, inhaling air or ingesting food contaminated by alpha radiation can be lethal. Beta particles can penetrate up to two centimeters of flesh and cause burns. Beta burns often take weeks to show up, however. A reddening of the skin called *erythema* usually appears and then goes away before the true severity of beta radiation injuries becomes apparent. Gamma rays are the most harmful: they can penetrate concrete and can also destroy bone marrow as they pass through the human body. Gamma radiation was the cause of death for many of Chernobyl's earliest victims.

Soon after the accident, 299 people were hospitalized for symptoms of radiation sickness. Of this group, which consisted mainly of plant operators and firefighters, 134 had confirmed cases of acute radiation syndrome (ARS), or radiation sickness. Acute radiation syndrome is a collection of symptoms resulting from radiation's effects on the body. These symptoms include a suppressed immune system caused by damage to bone marrow, digestive problems related to destruction of the lining of the intestinal tract, and skin damage from radiation burns. Hair loss, damage to the central nervous system, and radiation-induced hepatitis (inflammation of the liver) can also occur.

There are four degrees that describe the severity of radiation sickness. First-degree ARS is the least severe (the chance of survival is excellent with medical attention), while the prognosis of someone with fourth-degree ARS is very poor. Some of the sickest Chernobyl patients had combined radiation injuries—beta burns combined with thermal burns from the fires—that proved especially hard to treat. Within three months, 28 of the 134 people with ARS had died. The casualties were all power plant workers and firefighters. Twenty of them had fourth-degree ARS, seven had third-degree ARS, and one had second-degree radiation sickness.

When accident victims started arriving at the Pripyat medical center before dawn on April 26, doctors tried to categorize them by estimating their individual radiation doses. They based their estimates on the depth of each victim's nuclear tan, since even people with severe ARS feel better for a time after their initial vomiting subsides. That evening, physicians in Pripyat put the patients with the darkest nuclear tans on a plane bound for Moscow. Early on Sunday, April 27, this first group of 28 Chernobyl victims arrived at Moscow's Hospital No. Six, a medical center equipped to handle extensive radiation injuries. A second group of patients arrived later that day.

The USSR had refused offers of aid from the U.S. government, but the Soviets welcomed help from wealthy American industrialist and retired physician Armand Hammer. At his own expense, Hammer sent Dr. Robert P. Gale to Moscow to assist with the sickest patients. The 40-year-old Gale, a well-known authority from UCLA on bone marrow diseases and transplants, was joined by two UCLA colleagues and an Israeli doctor. The team worked with Dr. Angelina Guskova, who was in charge of treating radiation injuries at Hospital No. Six. The doctors wanted to test the usefulness of bone marrow and fetal liver tissue transplants as last-chance treatments for Chernobyl victims who would otherwise die.

A healthy person's bone marrow produces red blood cells that carry oxygen. Marrow also assists in blood clotting and produces white blood cells, called lymphocytes, which fight off infections. A small subgroup of the Chernobyl victims had absorbed as much as 10,000 times the amount of radiation most people would absorb in one year. They were gravely ill and extremely vulnerable to hemorrhages and infections. Dr. Guskova, Dr. Gale, and the rest of the transplant team chose 19 such patients to undergo experimental bone marrow or fetal liver tissue transplants.

The sickest patients received liver tissue because their immune systems were so ravaged that it was impossible to match them up with bone marrow donors. The transplant team injected donor marrow into the rest of the patients' bloodstreams. The patients with the best marrow matches were those who received their siblings' life-giving tissue. Others received donor marrow that had been treated to make it as compatible with their bodies as possible.

All of the fetal liver tissue recipients died, and just two of the bone marrow transplant patients lived. Both had sibling donors. Why did so many transplants fail? The doctors learned that many severely irradiated patients still had a miniscule number of immune cells circulating in their bodies—enough for the host's damaged immune system to rise up and attack the donor cells. Even the most suppressed immune systems were surprisingly resilient as they rallied against "foreign" liver and marrow tissues. Their experience with the Chernobyl victims taught the doctors that transplants are not a cure-all for severe radiation injuries.

This new knowledge would influence the way doctors and hospitals developed plans for coping with nuclear disasters or warfare. Researchers sought alternatives to marrow and fetal liver tissue transplants to treat radiation injuries. Today, drugs known as hemopoietic (blood-forming) growth factors can help patients regenerate their own bone marrow.

Dr. Robert Gale got celebrity treatment in the USSR

and public thanks from Mikhail Gorbachev. Television cameras recorded him strolling through Kiev with his family as if to suggest that the Soviet public had nothing to fear from radioactive contamination.

Once the initial crisis among firefighters and operators with acute radiation injuries and sickness ended, much of the world got over the shock of Chernobyl and went about the business of living. But for the exposed populations, long-term health problems posed an ominous threat. The human body cannot distinguish between radioactive particles and stable ones. If someone ingests, inhales, or absorbs radionuclides through the skin, they enter the bloodstream. Some behave like stable material inside the body, mimicking nutrients and minerals. Of the radionuclides in the Chernobyl plume the following were of particular concern to experts:

1. Cesium 137. It acts like potassium inside the body, targeting the gastrointestinal tract and muscle tissue and possibly speeding the aging process and increasing cancer risk.

2. Strontium 90. This isotope mimics calcium and is absorbed by bone, where it can damage marrow.

3. Iodine 131. It is rapidly absorbed by the thyroid gland, which can't distinguish it from stable iodine, and often causes thyroid cancer.

4. Plutonium 239 and other plutonium isotopes. These are absorbed by the lungs, the liver, and bones. They cause lung cancer in sufficient inhaled doses and may cause other cancers and cardiovascular illnesses.

Once inside the body, radionuclides continue to emit radiation until they decay into a stable material (this may take anywhere from seconds to millions of years) or leave the body via its wastes. They alter cells by bombarding them with electromagnetic energy. If the immune system fails to repair damaged cells before they proliferate, the result can be cancer.

Cancer can occur decades after radiation exposure—or it may never occur at all. Since the accident at Chernobyl was unprecedented, all scientists today can do is wait for its health effects to emerge. One often hears that Chernobyl was hundreds of times "worse" than the atomic bombings at Hiroshima and Nagasaki in 1945. This comparison is not very meaningful, however. Although the Chernobyl accident resulted in the gradual release of more radioactive material, the bombings of Nagasaki and Hiroshima resulted in immediate releases of high-level radiation that killed about 70,000 and 140,000 persons, respectively, within five months (that number includes people within close range of the bombs who were instantly vaporized by intense heat). Chernobyl's victims, on the other hand, have absorbed lower levels of radiation for prolonged periods of time.

Still, scientists used what they knew about the health effects of radiation in atomic bomb survivors to make some long-term health forecasts for people affected by Chernobyl. Radiation is known to trigger certain cancers years after exposure. It is unclear how much radiation a person needs to absorb before this happens, however. Furthermore, a few exposed people will develop cancer for other reasons (heavy smoking or chemical pollution, for example) while many others will remain cancer-free. Researchers nevertheless expected to see two clear cancer trends among the Chernobyl survivors within a few years of the accident: increases in thyroid cancer and leukemia (cancer of the blood).

Nine-year-old Yulia Kostina, from the southern Ukrainian Crimean Peninsula, is one of the region's many victims of thyroid cancer. After World War II, there was a recorded increase of thyroid cancer among Japanese survivors of the atomic bombs dropped on Hiroshima and Nagasaki. Health experts subsequently anticipated this same epidemic after the Chernobyl disaster, especially in younger segments of the population.

Studies of atomic bomb survivors showed a modest increase in the number of thyroid cancers about five years after the bombings. Experts watched and waited for an increase in thyroid cancers among Chernobyl survivors. They did not have to wait long. According to the World Health Organization (WHO), the number of thyroid cancer cases started going up in 1990. The thyroid glands of babies, children, and teenagers tend to absorb more radioactive iodine than those of adults. Therefore, the most shocking increases in thyroid cancer occurred among young people. Belarus, Ukraine's neighbor to the north, was the area most affected by the radioactive plume after the accident. From 1974 to 1985, just eight children in all of Belarus had thyroid cancer. During the years 1986–1997, that figure shot up to 574! Of those 574 Belarusian pediatric thyroid cancer patients, 305 of them lived in the Gomel district. It had rained in Gomel when the radioactive cloud was overhead, depositing huge amounts of iodine 131 in the air and soil.

Widespread distribution of stable iodine immediately after the accident could have spared many Belarusian children from thyroid cancer. For example, Poland was one of over 20 countries outside the USSR to receive fallout from Chernobyl. But the Polish government made an organized effort to give children potassium iodide when news of the accident broke, protecting them from thyroid cancer. Unfortunately, nobody can accurately predict when the rising number of thyroid cancers in Ukraine, Belarus, and parts of Russia will level off.

The news about leukemia cases after the accident is inconclusive. Although some doctors in the former USSR and eastern Europe report increased cases of leukemia since the accident, this increase has yet to be proven by hard data. This doesn't necessarily mean that leukemia cases aren't in fact increasing, or that they won't in the future.

If unborn children absorb enough radiation in the uterus, they can suffer from mental retardation. Children born to evacuee mothers or in contaminated areas haven't clearly demonstrated an increase in radiation-induced mental retardation. High radiation doses—either all at once or over time—do cause genetic changes in humans and animals. In 2000, Vladislav Ostapenko, an authority on radiation medicine in

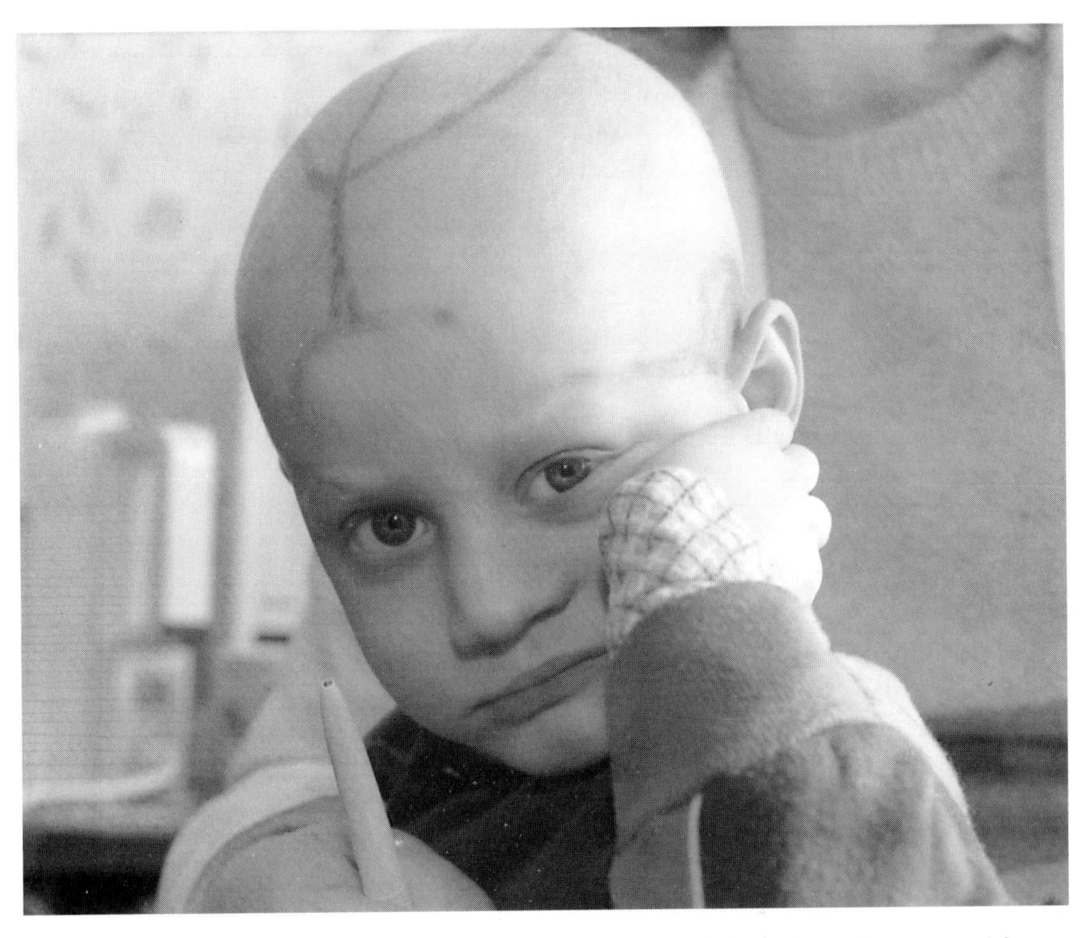

Alez Zhloba, aged five, suffers from leukemia, another type of cancer suspected to be associated with radiation exposure. Scientists do not have any hard data proving a direct correlation between the Chernobyl disaster and cases of leukemia, although doctors have reported an increase in the disease since the 1986 accident.

Belarus, blamed genetic changes caused by Chernobyl fallout for reproductive problems in young adults. He said that girls now showed five times more deformities in their reproductive systems than normal, and boys three times as many. "Science cannot yet assess the consequences of the Chernobyl accident, but it is plain that a demographic catastrophe has occurred in Belarus," he stated. A study reported in 2001 showed that children born to Ukrainian liquidators after the accident have an unusually high number of mutations in their DNA.

Whether these genetic changes will produce defects in future generations is unknown. Other effects of the Chernobyl accident are difficult to pin down in terms of their exact causes. Fatigue, sleep disturbances, and generalized feelings of illness that plagued many atomic bomb survivors in Hiroshima and Nagasaki were called "A-bomb disease." Many survivors of Chernobyl experience similar symptoms. Some authorities attribute these symptoms to stress alone, but a number of doctors in the former Soviet Union and eastern Europe believe that "Chernobyl AIDS" is a real illness. This term refers to a collection of immune system problems emerging in people exposed to radiation from the accident. Chernobyl AIDS is not a viral infection like true AIDS, nor is it contagious. Chernobyl AIDS sufferers do appear to have compromised immune systems, though. Symptoms include fatigue and weakness and a deficiency of "natural killer cells," a type of white blood cell instrumental in fighting off cancer cells and viruses. Recovery from simple colds, for example, takes longer than normal and often paves the way to more severe illnesses. Chernobyl AIDS patients are also more likely to develop blood cancers and other types of cancer than people without this condition.

But scientists are also quick to point out that Chernobyl may not be the only factor to blame if some people in the former Soviet republics suffer from weakened immune responses. The USSR had high levels of industrial pollution prior to the Chernobyl accident. Smoking was a common habit then and still is today. The increase of alcohol consumption by some evacuees and liquidators—as well as inadequate diets resulting from widespread poverty—may also play a role in the health problems observed among people affected by the Chernobyl accident.

There is no doubt, however, that the accident was a source of tremendous psychological stress for the evacuees, the liquidators, and other exposed people. For many, this stress continues today. People were traumatized when they discovered the true scale of the disaster and how their government had tried to conceal it. Some also lost the support of their friends and extended families through relocation. Some survivors' feelings of illness and fatigue in the absence of any known disease have been dismissed as "radiophobia." Studies have shown that survivors tend to attribute all kinds of sicknesses to their radiation doses. This fearfulness now blights the lives of many young people who weren't even born at the time of the accident.

Psychological stress is nothing to take lightly. Without relief, people under severe, chronic stress can develop physical and mental illnesses—or even die. Some researchers believe that stress is taking its toll on some of Chernobyl's evacuees and liquidators. Another group of special concern is the population that must remain in contaminated areas. These people have no way of knowing whether their long-term exposure will manifest itself in health problems—mainly leukemias and cancers—during the decades to come. Experts have referred to continuing studies of Hiroshima and Nagasaki survivors to predict that the cancer toll among Chernobyl survivors will probably increase by the 2030s.

On April 26, 2001, the World Health Organization estimated that some 30,000 deaths had occurred as a result of exposure to radiation released by the Chernobyl accident. Up to seven million people had been irradiated and/or seriously traumatized by the accident. The WHO released these figures on the accident's 15th anniversary to point out that victims in Ukraine, Belarus, and Russia in particular still need help. The governments of these former Soviet republics are in economic crises. Their

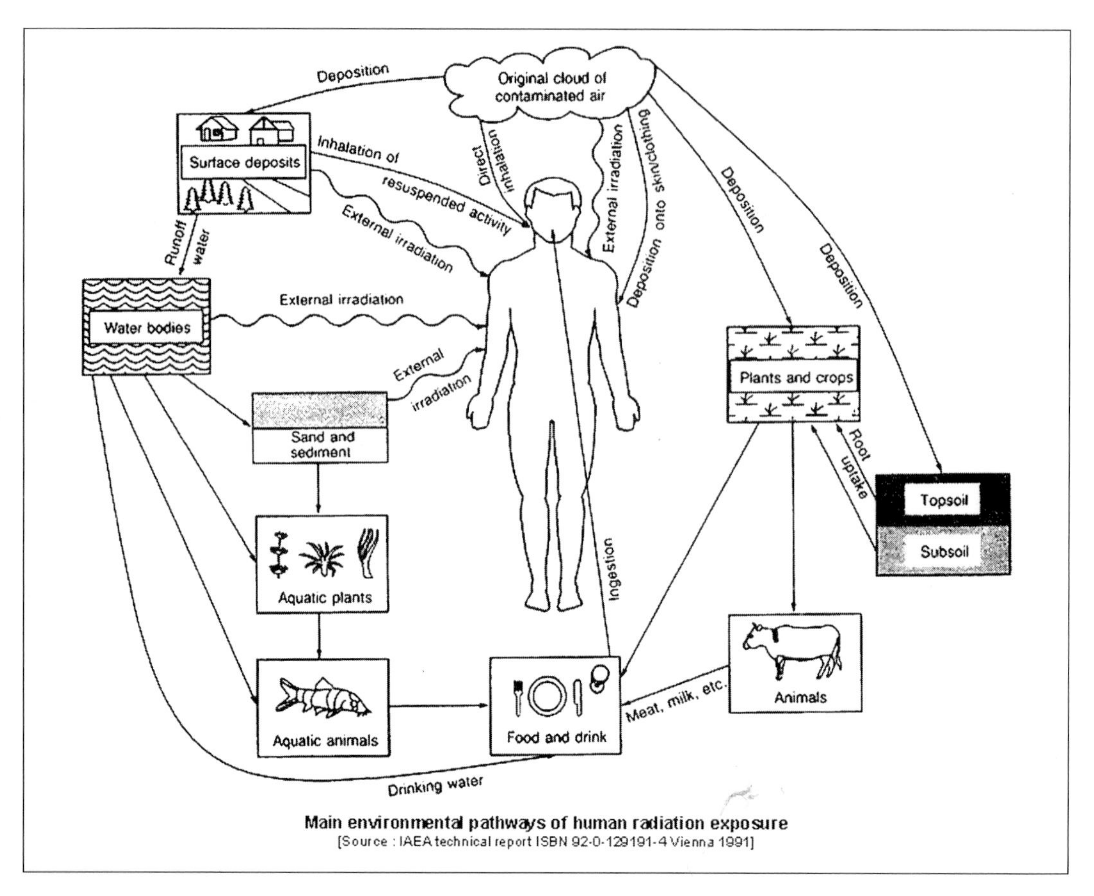


This diagram illustrates the many pathways between the environment and a human being susceptible to radiation. Scientists list five basic possible paths, all of which could be traced in radiation cases associated with Chernobyl fallout.

national budgets are stretched to the limit providing special benefits such as free housing and health care to people officially designated as victims of Chernobyl.

Perhaps no group has suffered more from the effects of Chernobyl than the liquidators. Some were soldiers conscripted to clean up the area around Chernobyl immediately after the accident; others were civilian volunteers. After the accident, some people alleged that men were recruited right off the street and compelled to work on the accident site. Researchers in Ukraine, with funding from the United States, have recently attempted to reconstruct the doses of radiation received by a sample of 135 liquidators by studying their teeth (tooth enamel can indicate a person's lifetime radiation dose).

This information may lead to valuable knowledge about how much radiation exposure it takes to affect people's health.

Studying the liquidators is problematic, however, because nobody is sure exactly how many of them served at Chernobyl. Liquidators are defined as cleanup and medical workers who helped mitigate the effects of the Chernobyl accident within the exclusion zone sometime during the years 1986–1991. Since the dissolution of the Soviet government in 1991, each of the 15 former Soviet Socialist Republics has kept its own records on its own liquidator population with varying degrees of accuracy. Estimates of their total numbers range from as many as 600,000 to 800,000 as stated at an international conference on Chernobyl in 1996, to as few as 168,000 according to Russia's National Dosimetric Registry. The possibility that some people falsely claim liquidator status in order to collect government benefits further complicates the issue.

What is known about the incomplete sample of certified liquidators is that many suffer from breathing problems such as chronic bronchitis and asthma. Male liquidators who received very high doses became incapable of producing children. The liquidators—many of whom are probably entering middle age today—suffer from a high incidence of heart and lung problems usually associated with elderly people. Some experience faintness upon exertion, accompanied by very high or very low blood pressure. The liquidators also seem to suffer from higher incidences of certain cancers, psychiatric illnesses, and deaths by suicide than the general population.

In 1989, a group of frightened and outraged liquidators banded together to demand more answers about radiation effects from the Soviet government and to get compensation for health problems they believe resulted from their cleanup work at Chernobyl. Headquartered in Kiev, the Chornobyl Union (Ukrainian spelling of "Chernobyl") is an active political force, with chapters all over the former Soviet Union.

On October 24, 2000, a group of Russian Chernobyl liquidators marched from Tula in central Russia to Moscow, protesting a law passed by the Russian Duma (parliament) that they believed would unnecessarily delay their receipt of government benefits. The protestors planned to start a hunger strike two days later.

Every anniversary of the Chernobyl accident brings dramatic public demonstrations by antinuclear groups and liquidator unions fighting to retain their benefits. The liquidators believe they are dying off unnaturally fast. Like almost every other figure pertaining to the human costs of history's worst nuclear power plant accident, the estimated number of Chernobyl-related deaths varies. Some of nuclear power's most ardent supporters say the toll will be negligible; some of its fiercest opponents say half a million or more will develop radiation-induced cancer as a result of the disaster in Ukraine.

The final toll Chernobyl will exact on human life remains unknown. Not enough was known about the pre-accident health of populations most affected by the radioactive plume to attribute their present illnesses to radiation alone. Many individuals may have developed poor health even if the accident hadn't happened because of inadequate diet, industrial pollution, or a lack of good medical care. There is also disagreement about exactly how much radiation was released into the environment after Chernobyl-4 blew apart. If it is more than the 50 million *curies* estimated by Soviet officials, then perhaps there will be more health problems in the future than anyone has previously predicted.

For the first few years after the accident, Soviet doctors were not allowed to attribute any health problems of their patients to Chernobyl. This also makes tracking the accident's true human toll difficult. The uncertainty surrounding the accident's impact is terrifying to the people affected by it and frustrating to the health professionals who now monitor Chernobyl victims. The future of the air, soil, and water around the NPP is similarly uncertain.

The enormous number of unknowns still haunting millions of people in Chernobyl's wake prompted many nations to renounce atomic energy—and specifically, the Soviet nuclear power program. Chernobyl would have to be shut down, many insisted. Would Ukraine take this step? If so, would the end of Chernobyl be enough to ensure that nothing like the firestorm in Reactor Four would occur again?

# The Fallout Continues

The long recovery after the Chernobyl disaster coincided with the fall of the Soviet Union's Communist Party. Rallies for Ukrainian independence from the USSR, like this one in August 1991, preceded the Ukrainian Parliament's official declaration of its independence in December of that year. Several of the candidates included among their campaign promises the permanent closing of Chernobyl.

The Chernobyl accident forced Soviet nuclear engineers to acknowledge serious safety deficiencies in the RBMK reactor design. These dangerous reactor units needed upgrades to decrease their positive void coefficient, to decrease the time required to scram, or shut down, the reactors in emergencies, and to make disabling their protective systems more difficult.

Engineers permanently installed 80 to 90 fixed control rods called absorber rods in each RBMK to slow down fission at low power levels. Their uranium fuel was enriched from below 2% to 2.4% to help keep the reactors stable and controllable at low power. Before the accident, it took 18 seconds to fully insert the control rods into an RBMK reactor core to shut it down; after post-Chernobyl safety improvements it took just 12 seconds. An emergency fast-scram system was also added to the controls. In addition, the required minimum of 30 control rods inserted in the reactor core at all times was increased to 43–48, depending upon the individual reactor. In addition, the control rods were redesigned to prevent a power surge when their tips entered the reactor core.

Operators at Chernobyl-4 had easily turned off the emergency core cooling system on April 25, 1986, to avoid ruining their turbogenerator test. To help prevent such foolish errors in judgment in the future, each RBMK reactor's ECCS was fitted with a two-key access system to prevent any one operator from disabling it without consulting others.

The world saw these mechanical improvements as temporary stopgaps until safer reactors could replace the Soviet RBMKs or—as many hoped—until the USSR halted its nuclear power program. While the Communist Party's enthusiasm for atomic energy never wavered because of Chernobyl, many individuals within the Soviet Union felt differently. The accident helped start a widespread societal shift that put an almost 70-year-old government system to the test. The Chernobyl accident coincided with *perestroika*, a plan instituted by Secretary General Mikhail Gorbachev to change the Soviet Union's entrenched economic and political structure. Gorbachev was also fostering glasnost, which created a freer exchange of information between the USSR and other nations, as well as more access to accurate information for ordinary Soviet people.

When Chernobyl-4 tore open and spewed its radioactive cargo into the sky, several unsafe reactor units were either in the planning stages or under construction in the Soviet Union. Right after the accident, the Kremlin—the seat of the Soviet government—

After years of petitions by the United States and other major countries, Chernobyl-3, the last working reactor of the power plant, was finally shut down on December 15, 2000.

pledged to continue building more nuclear power plants in the USSR. But the spirit of glasnost had motivated ordinary Soviet citizens, many of whom now did not want nuclear reactors in their towns. The accident prompted people who had never before dared to oppose the repressive Communist Party to demonstrate against nuclear power.

As a result of the Chernobyl accident, all plans for new RBMK reactors were soon scrapped. There were five such reactors under construction in the USSR. People successfully protested their completion—even with the promise of new safety modifications—without fear of being punished by the Communist Party. Early in 1988, the Soviet government abandoned a partially built NPP at Krasnodar in Russia, absorbing $43 million of wasted funds. There was nothing structurally wrong with the Krasnodar plant—people just didn't want it in their town.

In addition, new pressurized water reactors in the Ukrainian towns of Khmelnitsky and Rovno were going up at the time of the Chernobyl accident. Although these reactor units are safer than RBMKs, they still fall far short of Western safety standards. Citizens of Khmelnitsky and Rovno opposed the completion of the two reactors. The USSR abandoned construction of Khemlnitsky-2 and Rovno-4 in response to public pressure.

From the earliest post-accident days, Soviet print media had also shown unprecedented boldness on nuclear power issues. Articles candidly criticized the Communist Party's attempts to cover up the true scale of the Chernobyl disaster. The spirit of glasnost also inspired artistic expression. Vladimir Gubaryev, science editor of the Russian magazine *Pravda*, wrote a play called *Sarcophagus*. The play takes place in a radiation research hospital as it receives victims of the recent Chernobyl accident. As the fictitious plant director lies dying of radiation injuries, "K," the play's main character—who survives despite a "fatal" dose of radiation during a lab experiment a year earlier—offers to donate his bone marrow. K's explanation for his generosity points an accusatory finger at the real plant director, Viktor P. Bryukhanov (who was not injured in the explosion): "I want to condemn him to life," K says. Remarkably, Gubaryev's controversial play appeared in the September 1986 issue of the Soviet literary magazine

*Znamya*, and its author faced no reprisals from the government.

But not everyone who spoke out publicly about Chernobyl went unscathed. Vladimir Gubaryev invited Valery Legasov, deputy director of the Kurchatov Institute of Atomic Energy, to contribute a piece on the Chernobyl disaster to *Pravda*. The resulting article, with a title that translated reads, "My Duty Is to Tell about This . . .," was a revelation. Legasov had worked closely with Anatoly Alexandrov to design the first commercial RBMK reactors. After the accident, it was Legasov who headed the Soviet delegation to Vienna in the summer of 1986 and defended his nation's nuclear power program to the world by blaming the explosion on the foolish actions of the operators. In *Pravda*, however, Legasov disclosed his true belief that the lack of safety standards at Chernobyl was not entirely their fault. He pointed out that they were only following orders—and unclear orders at that:

> I have a record of the telephone calls of the operators just before the accident and it is terrible to read them. One operator calls to another: "In the programme [sic] it is written what to do and then many points are crossed out, what should I do?" Another operator thinks for a few seconds and replies: "Do those things which are crossed out."

Legasov also wrote of his personal concerns about the inherent dangers of RBMK reactors. "As a chemist," he wrote, "I was worried about the amount of graphite, zirconium and water which was used in the design of the RBMK. I also considered that the design of the reactor was unusual and inefficient from the point of view of protective systems which should come into operation under extreme conditions."

Until Chernobyl, Legasov's scientific career was headed for a brilliant finish. It seemed inevitable that he would soon replace Anatoly Alexandrov as director of the Kurchatov Institute. But when he stopped mouthing the Communist Party line and started to openly question long-established nuclear power policy and procedures of the USSR, his career stalled. Party officials and Legasov's fellow scientists increasingly ignored his input on atomic energy matters. "My Duty Is to Tell about This . . ." appeared in *Pravda* on May 20, 1988, but Legasov did not live to see it in print; he had taken his own life on April 25, 1988, just one day shy of the second anniversary of Chernobyl. His proposal to overhaul safety standards in the nuclear power industry was defeated the day before his death. Soviet officials attributed Legasov's suicide to depression over having been diagnosed with leukemia. His widow, Margarita, disputed this explanation. She maintains that her husband was despondent because the Communist Party ruined his career to punish him for his candor about deficiencies in the Soviet nuclear power program.

The life and death of Valery Legasov parallels a change in the collective mindset of many people in the Soviet Union at the end of the 1980s. In 1989, the faltering USSR withdrew from Afghanistan, an Asian country about the size of Texas, after an attempted Communist takeover that lasted 10 years. The Soviet republics wanted independence from Russia, and ordinary people were growing tired of the Party's use of repression, fear, and misinformation to enforce loyalty to the government. As general secretary of the Communist Party, Mikhail Gorbachev had begun to loosen the Party's grip on media outlets, encouraging more factual information and less propaganda in print and on radio and TV. Gorbachev also envisioned a market economy rather than Communism, and elections in which people could choose from multiple candidates. In 1991, his successor Boris Yeltsin became Russia's first freely elected president.

Despite a failed attempt by Party hard-liners to prevent the fall of Communism, the Union of Soviet Socialist Republics dissolved. On December 1, 1991, the Ukrainian Parliament declared its independence from the USSR. The 15 former Soviet republics became the loosely defined Russian Federation. The Communist Party's secrecy after the Chernobyl accident undoubtedly hastened the demise of the USSR. The first popularly elected Ukrainian parliament had several candidates who had included closing down Chernobyl among their campaign promises. The Chernobyl accident was also instrumental in the rise of the environmental-activist Green Party in Ukraine.

Chernobyl-2 was shut down for good after a turbine hall fire in 1991. Chernobyl-4 was, of course, already permanently off-line as a result of the world's worst technological disaster. It sits encased inside an aging 10-story concrete and metal structure commonly known as the Sarcophagus. Chernobyl-1 was taken off-line in late 1996, after pressure tubes inside the reactor unit became too old, according to their manufacturer's guidelines.

Chernobyl-3 remained on-line to generate electricity until a safer energy source could be found, or until Ukraine bowed to Western pressure to close the notorious plant. The economically stagnant Ukraine wanted a guarantee of monetary aid to *decommission* Chernobyl, to provide for more nuclear waste storage, and to complete two new nuclear reactors before agreeing to shut down the world's most notorious NPP. Specifically, Ukraine wanted to complete the construction of the previously abandoned reactor units in Khmelnitsky and Rovno, two small cities west of the Chernobyl plant. Khmelnitsky-2 and Rovno-4 were both Soviet-designed pressurized water reactors with safety problems, although they were less dangerous than the RBMK type. The so-called G7 (Group of Seven countries: the United States, Canada, the United Kingdom, Germany, France, Italy, and Japan) and the European Union (EU) initially wanted Ukraine to use any aid it received to upgrade its traditional coal- and gas-fired electrical power plants. But this solution would have added to the level of greenhouse gases already in the environment. Moreover, Ukraine was already deeply in debt to Russia for natural gas, so phasing out nuclear power would have led to additional dependence upon Russia.

Many wondered why Ukraine would want to throw money at reactors that were so unsafe yet so near completion that they would be almost impossible to modify. Opponents of the plan argued that Ukraine did not need any power sources to replace Chernobyl, anyway. They claimed that impoverished Ukraine's demand for power had fallen so low that even without Chernobyl, there would be no need for new generators until 2010 at the earliest.

Ukraine, on the other hand, saw Khmelnitsky-2 and Rovno-4 as potential sources of abundant electricity and new jobs in a land of high unemployment. The prospect of new job openings was especially important to Ukraine because closing Chernobyl would eventually leave thousands of plant operators out of work. On December 20, 1995, Ukraine, the G7 countries, and the EU nonetheless signed the Memorandum of Understanding. In this document, Ukraine agreed to shut down Chernobyl-3 by 2000 in exchange for over $3 billion in aid.

Even after the signing of the Memorandum of Understanding, however, international squabbles continued. Ukrainian president Leonid Kuchma shut

Ukrainian president Leonid Kuchma yielded to other countries' pleas to close down Chernobyl for good, but at the same time he pushed for the construction of two other reactor units to replace those at Chernobyl. In 2000 the Ukraine agreed to shut down Chernobyl-3 in exchange for $3 billion in foreign aid to fund the building of Khmelnitsky-2 and Rovno-4.

down Chernobyl-1 in late 1996, fulfilling a promise he had made that April. By 2000, Chernobyl-3 alone supplied five percent of Ukraine's electricity. But Kuchma maintained that shutting down Chernobyl-3 in 2000 would be impossible without two new Ukrainian reactors to take its place on the power grid—or at least some financial aid in advance. The G7 and EU countries were just as adamant in their refusal to assist the Ukrainian government until it had a firm shutdown date for Chernobyl-3.

Although Khmelnitsky-2 and Rovno-4 remain unfinished at this writing, Kuchma ended the standoff by setting December 15, 2000, as the shutdown date for

Chernobyl-3, the plant's last working nuclear reactor. Former United States president Bill Clinton called the decision "a triumph for the common good." Kuchma took a less triumphal view of the shutdown, saying that Ukraine was "prepared to sacrifice part of its natural interest for the sake of global safety." As TV cameras bore silent witness, engineer Sergey Bashtovoi made history by flipping the shutdown switch on the control panel of Chernobyl-3 for the last time.

Not everybody felt like celebrating after the shutdown was completed, however. Until December 15, 2000, about 5,800 people still worked at the Chernobyl NPP. They lived in Slavutich, a new settlement built to replace Pripyat, located 30 km from the plant. The shutdown of Chernobyl meant they would soon be out of work, making ends meet in Ukraine's depressed economy. Their jobs would gradually dry up until about 2,000 remained to do cleanup work and to help fix the aging Sarcophagus that still houses Chernobyl-4.

Its proper name is the Ukritiye Encasement, but the crumbling 10-story structure housing the crippled reactor is better known as the Sarcophagus. Approximately 117,000 liquidators risked huge doses of radiation to build the Sarcophagus. Construction of the huge concrete and metal structure took place from May 18 to November 19, 1986—a remarkably short time considering the enormity of the liquidators' task.

The Communist Party assigned the Ministry of Mid-Level Industrial Engineering to build a shelter around Reactor Four in 1986, just days after the accident. Ministry department heads quickly selected one of 18 proposed designs and sent the liquidators to work. A group of liquidators had already tunneled underneath Chernobyl-4 to put a concrete cooling slab on the reactor unit floor. This slab prevented molten

This picture from inside the Chernobyl-4 reactor shows just a fragment of its total melted nuclear fuel, which is an estimated 10 metric tons. The Ukraine still faces the challenge of preserving the Sarcophagus, the structure built to contain the radioactive materials, so that no deadly particles escape into the air again.

core material from burning into the ground. Had the hot radioactive material melted into the underground water table, a massive explosion could have resulted.

The designers of the Sarcophagus expected it to last at least 30 years. Unfortunately, it is a product of the hurried and dangerous conditions under which it was constructed, though, and it is now deteriorating rapidly. The cement walls of the Sarcophagus have numerous cracks, permitting snow, dew, rainwater, and even birds to go deep into its radioactive belly.

From 1995 to 1997, the Sarcophagus underwent some repairs to extend its life, but these were only temporary fixes. Workers regularly treat the interior of the

Sarcophagus with a water-based dust suppressor since radioactive dust remains a serious threat around the entombed reactor unit.

An estimated 10 metric tons of radioactive core material lurks beneath the sand mixture used to smother the graphite fire in 1986. Should the Sarcophagus collapse, these core particles could become part of a dust cloud. The contents of a radioactive dust cloud could spread for hundreds of kilometers around the power plant. Radioactive water currently pooled on the Sarcophagus floor could reach the underground water table. Events that might cause a collapse include hurricanes, tornadoes, terrorist attacks (one plot was thwarted in 2000), and earthquakes. The Sarcophagus survived a magnitude-four quake in 1990. According to the Ukrainian Academy of Sciences, stronger tremors are possible within the exclusion zone.

Ukraine has long been aware of the Sarcophagus's structural problems. In 1992, engineers around the world competed to develop the best plan for containing the crumbling enclosure. A French design for a "supersarcophagus" won approval in 1994. The project will cost $300 million for construction alone, which *Pravda* says is scheduled for completion in 2007. That price tag does not include design costs or approximately 30 years of cleanup work that will be necessary at the accident site.

Since the collapse of the Soviet Union in 1991, the transition from communism to a free market economy has not been smooth; few people prosper, while many more struggle to feed and clothe their families. Ukraine, Belarus, and Russia still feel the financial strain of coping with Chernobyl. Charities from around the world try to help with gifts of money, medicines, physicians, and supplies to treat adults and children affected by the accident.

A chief surgeon from a children's hospital in Kiev poses with equipment and supplies donated by the James Robison Relief agency. The people of the former Soviet Union have faced a difficult transition to a free economy, and Ukrainian doctors still treating Chernobyl-related illnesses have had to rely on donations to keep their hospitals running.

Against this bleak economic backdrop, it is reasonable to ask, will Chernobyl stay closed? President Leonid Kuchma spoke of the Ukrainian people's sacrifice when the infamous Chernobyl NPP closed forever on December 15, 2000. Before the shutdown, residents of Kiev were already shivering through scheduled rolling blackouts to conserve energy.

Ukraine still may decide to refurbish its old RBMK reactors to run for decades to come. Currently, there are 11 RBMK reactors still operating in Russia and 2 in Lithuania. These reactors underwent safety upgrades in the wake of the Chernobyl accident, but they are still dangerous. They are also in poor repair because Russia cannot afford to maintain them well. In 1999,

*Newsweek* magazine got authorization to visit the Leningrad NPP. The facility is located in the Russian town of Sosnovy-Bor outside of St. Petersburg. Improvements to the three operational reactors at Leningrad were supposed to have taken place by late 1999. Those changes were tentatively rescheduled for 2001 at the time of *Newsweek*'s visit—and then only if the necessary funds materialized.

The correspondents and photographers found a surprising lack of safety standards inside the plant:

> Visitors to the LNPP remove their shoes and don plastic booties and protective overclothing, but this seems intended mainly for the amusement of the staff. In the Block 2 Reactor Containment Room, for instance, many employees don't wear any protective gear at all—or even dosimeters.

It is clear that even if the Chernobyl NPP remains off-line forever, other hazards abound in the former Soviet bloc countries. The crushing economic conditions there encourage the continued use of obsolete nuclear reactors. "The Russian government," wrote *Time Europe* correspondent James O. Jackson in September 2000, "has figured out that its nuclear generators—lacking the sophistication and elaborate safety features imposed on Western nukes—can produce electricity more cheaply than gas- or coal-fired stations." Jackson explained that because of this, Russia is moving to increase its own reliance on nuclear power while Gazprom, a government-run enterprise, sells natural gas to Western consumers. This scheme may help Russia's ailing economy in the short term; the long-term effects of keeping Soviet-designed NPPs on-line may be disastrous, however.

Ironically, Russia's plan to stage an economic recovery by exporting its plentiful natural gas while relying on nuclear fission for its own needs may succeed precisely because many Western nations have turned away from nuclear power—in response to the Chernobyl accident!

The aftermath of Chernobyl continues, as it will for generations to come. How much reform has taken place? Numerous national and international oversight committees have been impaneled in the years since the world's worst nuclear power plant accident terrified the world. The hows, whys, and what-ifs of Chernobyl have spawned an entire field of study devoted to this single catastrophe. The people who lived around the Chernobyl NPP on April 26, 1986, face lifelong medical and scientific follow-ups.

Some of the changes brought about by the accident were positive; people were more emboldened to question their totalitarian government and prompt the nuclear power industry to reevaluate old reactor designs and safety practices. As *Newsweek*'s 1999 visit to the Leningrad NPP showed, the transformation is hardly complete. But there is ample room for optimism. The Soviet nuclear power program was rife with cover-ups and devoid of a healthy respect for the unchained atom's destructive power. The accident made people question their country's isolationism and repression of information. When Chernobyl-4 spat its concrete lid into the air and tens of millions of curies of radiation along with it, the USSR's stony silence seemed to break open a little, too. After Chernobyl, Soviet nuclear power plant officials visited Western facilities, and Western plant directors were allowed to penetrate the veil of secrecy surrounding Soviet plants. In 1995, Ukraine ratified the Law on

At the 15th anniversary of the 1986 disaster, Ukrainian president Leonid Kuchma stands in front of a monument honoring the Chernobyl cleanup workers who have died.

Atomic Energy Utilization and Radiation Safety. Its most important clause states that safety is now the first priority in all uses of atomic energy.

The effects of Chernobyl are evident in the painful sight of suburban Moscow's Mitino Cemetery, where lead-lined coffins soldered shut hold those firefighters and operators first to die at the disaster, as if they had become mere nuclear waste in death. Yet the inevitable recovery is illustrated by an awkward but lighthearted moment in "Back to Chernobyl," a 1989 installment of the American public television series *Nova*. In the show, Dr. Richard Wilson, an American nuclear physics expert, visits the Chernobyl plant and presents Mikhail Umanets, the plant's new director, with a gag gift. It is a T-shirt that reads, "Nuclear Energy: Safer than Sex." After an interpreter translates the slogan, Umanets laughs heartily, an unusual reaction in a place where laughter had been unthinkable such a short time earlier.

Few people associate the word "Chernobyl" with anything positive. In the long run, though, the disaster in Ukraine may help temper our technological arrogance with a degree of caution. If the lesson remains fresh even as time relegates Chernobyl to the back pages of our newspapers and the recesses of our minds, then the invisible poisons of Reactor Four will not have fallen to earth in vain.

# Chronology

1979 Human error and mechanical problems cause partial core meltdown at the Three Mile Island NPP in Middletown, Pennsylvania, on March 28

1983 Chernobyl-4 starts up on December 20

1985 The Soviet Union's Ministry of Energy forbids mass media from reporting about the "unfavorable ecological impact of energy-related facilities" on May 19; explosion and released steam kills 14 people in accident during startup of a reactor unit at the Balakovo NPP in Russia on June 27

1986 *April 26:* Steam and hydrogen explosions destroy Reactor Four at the Chernobyl NPP in Ukraine in the worst peacetime nuclear accident to date

*April 28:* Officers at the Forsmark NPP in Sweden detect high radiation levels coming from outside their plant; the Soviet government acknowledges the Chernobyl accident

*August:* Delegates from 62 countries meet in Vienna, Austria, to hear full Soviet report about the Chernobyl disaster during meeting of the International Atomic Energy Agency (IAEA)

1987 In July, six defendants are tried and found guilty of charges related to the Chernobyl accident

1988 In response to citizens' protests, the Soviet government abandons the Krasnodar NPP in Russia before construction is completed; Valery Legasov commits suicide on April 25, weeks before his writings about design flaws of the RBMK reactor and lax safety practices at pre-accident Chernobyl appear in the magazine *Pravda*

1991 Soviet Union collapses; Ukraine, Belarus, Russia, and other former Soviet republics become independent

1995 Ukraine and the G7 countries sign the Memorandum of Understanding on December 20, in which Ukraine agrees to close the Chernobyl NPP by 2000 in exchange for monetary aid

1999 Accident at nuclear fuel fabrication plant in Tokaimura, Japan, on September 30 kills one and injures two other workers. Dozens of others are exposed to high radiation levels

2000 Chernobyl-3, the plant's last working reactor, is shut down on December 15

2001 Study shows that children of Ukrainian liquidators display genetic mutations in their DNA; on April 26, the 15th anniversary of the disaster, the World Health Organization estimates that 30,000 deaths can be attributed to the Chernobyl disaster and that seven million people have been exposed to radiation from the accident

# Glossary

**chain reaction:** in nuclear physics, a self-sustaining fission reaction

**control rods:** rods made of neutron-absorbing material such as boron that can be lowered into a nuclear reactor's core to decrease power or shut down the reactor

**curie:** the amount of radioactive material in which 37 billion atoms per second undergo radioactive decay

**decommission:** to permanently close down a nuclear reactor

**dosimeter:** a device that measures an absorbed dose of ionizing radiation

**erythema:** reddening of the skin that precedes the appearance of burns from beta radiation

**fission:** the release of energy through the splitting of an atom's nucleus by a neutron

**glasnost:** a policy of more open distribution of information by the government of the former Soviet Union

**half-life:** the amount of time it takes for one-half of a radioactive substance's atoms to decay

**isotope:** any of several subtypes of certain elements whose atoms are similar to each other but have slightly different physical properties

**liquidators:** workers who volunteered or whom the government conscripted to get rid of the consequences of the Chernobyl accident

**meltdown:** the accidental melting of a nuclear reactor's core

**moderator:** a material that regulates the speed of nuclear fission so that a chain reaction is sustainable

**nucleus (pl. nuclei):** The center of an atom, composed of positively charged protons and uncharged neutrons

**neutron:** an uncharged particle found in the nucleus of nearly every element

# Glossary

**perestroika:** the plan instituted during the mid-1980s to reconfigure the economy and political structure of the former Soviet Union

**radiation:** energy in the form of electromagnetic waves or moving particles

**radioactive:** emitting highly dangerous particles and waves of radiation

**radionuclides:** elements that give off radiation

**roentgen:** a unit that measures exposure to ionizing radiation (X rays or gamma rays); named after the German scientist Wilhelm Roentgen, who discovered X rays in 1895

**unstable:** a characteristic of radioactive isotopes whose nuclei constantly decay until they eventually become stable isotopes

# Further Reading

**Books and Articles**


Arnaout, Ramy A. "Sich Discovers Chernobyl Worse than Prior Reports." *The Tech,* February 1, 1994.

Cheney, Glenn Allen. *Chernobyl.* New York: New Discovery Books, 1993.

Greenwald, John. "Judgment at Chernobyl." *Time,* July 20, 1987.

Gubaryev, Vladimir. Trans. Michael Glenny. *Sarcophagus.* Dramatic Publishing Co., 1990.

Jackson, James O. "Dirty Rotten Reactors." *Time Europe,* September 18, 2000.

Lampton, Christopher F. *Nuclear Accident.* Brookfield, Conn.: Millbrook Press, 1994.

Lemonick, Michael D. "The Chernobyl Cover-up." *Time,* November 13, 1989.

Medvedev, Grigory. Trans. Evelyn Rossiter. *The Truth about Chernobyl.* U.S.A.: Basic Books, Inc., 1991.

Medvedev, Zhores A. *The Legacy of Chernobyl.* New York: W.W. Norton and Co., 1992.

Mould, Richard F. *Chernobyl Record.* Philadelphia: Institute of Physics Publishing, 2000.

Nardo, Don. Illus. Brian McGovern. *World Disasters: Chernobyl.* San Diego: Lucent Books, Inc., 1990.

NEA Committee on Radiation Protection and Public Health. *Chernobyl Ten Years On: Radiological and Health Impact,* November 1995.

Nordland, Rod, Scott Johnson, and Stefan Theil. "Where Is the Next Chernobyl?" *Newsweek,* October 18, 1999.

Reuters News Service. "Post-Chernobyl Genetic Disaster in Belarus." March 1, 2000.

# Further Reading

Serrill, Michael S. "We Are Still Not Satisfied." *Time,* September 8, 1986.

Shcherbak, Yuri M. "Ten Years of the Chernobyl Era." *Scientific American,* April 1996.

Vargo, George J., ed. *The Chornobyl Accident: A Comprehensive Risk Assessment.* Columbus, Ohio: Battelle Press, 2000.

**Websites**

CBC News Online Article—"Children Born after Chernobyl Show DNA Changes." *http://cbc.ca/*

Facts About the Chernobyl Disaster
*http://www.belarusguide.com/chernobyl/chfacts.htm*

The Nuclear Control Institute
*http://www.nci.org/index.htm*

PBS Science and Technology—"Frontline: Nuclear Reaction"
*http://www.pbs.org/wgbh/pages/frontline/shows/reaction*

Pravda.RU Online Article—"Fifteen Years Have Passed Since the Chernobyl Accident."
http://english.pravda.ru/*society/2001/04/26/4282.html*

# Index

# Index

# Index

# Picture Credits

page

2: AFP/Corbis
10: Associated Press, AP
14: Megellan Geographix/ Corbis
17: Associated Press, AP
20: Associated Press, AP
22: Associated Press, AP
26: OECD, material available at: http//www.nea.fr/html/rp/ chernobyl/chernz.gif
31: Associated Press, AP
36: AFP/Corbis
42: Courtesy of U.S. Department of Energy
45: Courtesy of U.S. Department of Energy
47: AFP/Corbis
48: Courtesy of U.S. Department of Energy
50: AFP/Corbis
54: Corbis
59: Associated Press, AP
62: Caroline Penn/Corbis
66: Wolfgang Kaehler/Corbis
68: APF/Corbis
69: Associated Press, AP
73: Associated Press, AP
74: Associated Press, AP
81: Associated Press, AP
83: Associated Press, AP
86: IAEA
90: David &Peter Turnley/ Corbis
93: Reuters NewMedia Inc./ Corbis
99: Associated Press, AP
101: Courtesy of U.S. Department of Energy
103: Jim McDonald/Corbis
106: Associated Press, Pool

*Front cover:* Associated Press, AP
*Back cover:* AFP/Corbis

KRISTINE BRENNAN is a writer and editor in the Philadelphia area, where she lives with her husband and sons. She holds a B.A. in English with a concentration in professional writing from Elizabethtown College. She is also the author of *The Stock Market Crash of 1929* in this series.

JILL McCAFFREY has served for four years as national chairman of the Armed Forces Emergency Services of the American Red Cross. Ms. McCaffrey also serves on the board of directors for Knollwood—the Army Distaff Hall. The former Jill Ann Faulkner, a Massachusetts native, is the wife of Barry R. McCaffrey, who served in President Bill Clinton's cabinet as director of the White House Office of National Drug Control Policy. The McCaffreys are the parents of three grown children: Sean, a major in the U.S. Army; Tara, an intensive care nurse and captain in the National Guard; and Amy, a seventh grade teacher. The McCaffreys also have two grandchildren, Michael and Jack.